Matteo Fischetti

Introduction
to
Mathematical Optimization

Introduction to Mathematical Optimization

Matteo Fischetti

September 12, 2019

Preface

This book is intended to be a teaching aid for students of the courses in *Operations Research* and *Mathematical Optimization* for scientific faculties. Some of the basic topics of Operations Research and Optimization will be considered: Linear Programming, Integer Linear Programming, Computational Complexity, and Graph Theory. Particular emphasis is given to Integer Linear Programming, with an exposition of the most recent resolution techniques, and in particular of the *branch-and-cut* method. The work is accompanied by numerous examples and exercises.

The text is taken from the notes of Ivan Brugiolo for the course in *Operations Research* I held, in the academic year 1994-1995, for the degree course in Computer Engineering of the University of Padua. The initial draft was subsequently revised and integrated.

The cover is an elaboration of a drawing made by my daughter Martina at the age of 7.

The TEX compiler, version Big C 3.1415 for WindowsNT^{TM}, with the LATEX 2_ε macro package, version 01/06/95, was used for the drafting of the text.

Padua, September 2019: Fifth edition, in English, with Kindle Direct Publishing as editor. I would like to thank Susanna Legnaro, who translated this book into English, and Domenico Salvagnin, who proofread it.

Matteo Fischetti

Contents

Notation

- $\Re$ = set of real numbers

- Z = set of integers (positive, negative or zero)

- $[a, b] = \{x \in \Re \, : \, a \leq x \leq b\}$ = closed interval

- $(a, b) = \{x \in \Re \, : \, a < x < b\}$ = open interval

- $\mathbf{x} = \begin{bmatrix} x_1 \\ \vdots \\ x_n \end{bmatrix}$ = n-dimensional column vector $(\mathbf{x} \in \Re^n)$

- $(\mathbf{x}, \mathbf{y}) = \begin{bmatrix} \mathbf{x} \\ \mathbf{y} \end{bmatrix}$ (used for typographical purposes)

- $\mathbf{c}^T = [c_1, \ldots, c_n]$ = n-dimensional row vector (T is the transpose operator)

- $\mathbf{c}^T \mathbf{x} = \sum_{j=1}^n c_j x_j$ = scalar product between the vectors $\mathbf{c}, \mathbf{x} \in \Re^n$

- $||\mathbf{x}|| = \sqrt{\sum_{j=1}^n x_j^2}$ = Euclidean norm of the vector $\mathbf{x} \in \Re^n$

- $A = \begin{bmatrix} a_{11} & \cdots & a_{1n} \\ \vdots & & \vdots \\ a_{m1} & \cdots & a_{mn} \end{bmatrix} = \begin{bmatrix} \mathbf{a}_1^T \\ \vdots \\ \mathbf{a}_m^T \end{bmatrix} = [A_1, \ldots, A_n]$ = $m \times n$ matrix

- $A\mathbf{x} = \sum_{j=1}^n A_j x_j$ = column vector obtained combining the n columns of A

- $\mathbf{u}^T A = \sum_{i=1}^m u_i \mathbf{a}_i^T$ = row vector obtained combining the m rows of A

- $\arg\min\{f(i) \, : \, i \in I\}$ = argument $i^* \in I$ such that $f(i^*) = \min\{f(i) \, : \, i \in I\}$

- $\lfloor z \rfloor = \max\{i \in Z \, : \, i \leq z\}$, defined for all $z \in \Re$

- $\lceil z \rceil = \min\{i \in Z \, : \, i \geq z\}$, defined for all $z \in \Re$

- $\varphi(z) = z - \lfloor z \rfloor \geq 0$ = fractional part of $z \in \Re$

- $|z|$ = absolute value of the scalar $z \in \Re$

- $|Q|$ = number of elements (cardinality) of set Q

Chapter 1

Mathematical Programming

A Mathematical Programming (or Mathematical Optimization) problem can be formulated as:

$$\begin{cases} \min f(\mathbf{x}) \\ \mathbf{x} \in X \end{cases}, \tag{1.1}$$

where $X \subseteq \Re^n$ is *the set of feasible solutions* and $f : X \to \Re$ is the *objective function*. The convention is to formulate each problem as a minimization problem and, where appropriate, operate the substitution $\max\{f(\mathbf{x}) : \mathbf{x} \in X\} = -\min\{-f(\mathbf{x}) : \mathbf{x} \in X\}$.

A problem for which there is no feasible solution ($X = \emptyset$) is said to be *infeasible*; in this case, we agree to write $\min\{f(\mathbf{x}) : \mathbf{x} \in X\} = +\infty$.

A problem for which f is not bounded from below in X is hence said to be an *unbounded* problem; in this case, we write $\min\{f(\mathbf{x}) : \mathbf{x} \in X\} = -\infty$.

Solving problem (1.1) consists in identifying an *optimal solution*, if any, i.e., a solution $\mathbf{x}^* \in X$ such that $f(\mathbf{x}^*) \leq f(\mathbf{x})$ for all $\mathbf{x} \in X$. This solution is not necessarily unique.

A solution $\overline{\mathbf{x}} \in X$ is said to be *locally optimal* if $\varepsilon > 0$ exists such that $f(\overline{\mathbf{x}}) \leq f(\mathbf{x})$ for all $\mathbf{x} \in X$ with $||\mathbf{x} - \overline{\mathbf{x}}|| \leq \varepsilon$.

1.1 Convex Programming

Definition 1.1.1 *Given* $\mathbf{x}, \mathbf{y} \in \Re^n$, *the point* $\mathbf{z} := \lambda\mathbf{x} + (1 - \lambda)\mathbf{y}$ *is said to be a* convex combination *of* $\mathbf{x}, \mathbf{y}$ *for all* $\lambda \in [0, 1]$. *The combination is said to be* strict *if* $0 < \lambda < 1$.

More generally, the convex combination of k points $\mathbf{x}^1, ..., \mathbf{x}^k \in \Re^n$ is defined as $\mathbf{z} := \sum_{i=1}^{k} \lambda_i \mathbf{x}^i$, with $\lambda_1, ..., \lambda_k \geq 0$ and $\sum_{i=1}^{k} \lambda_i = 1$.

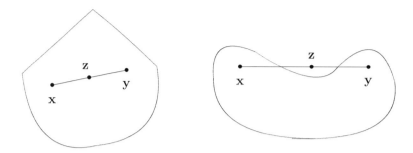

Figure 1.1: A convex set (on the left) and a non-convex set (on the right)

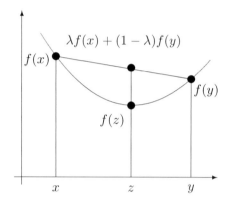

Figure 1.2: A convex function $f(x)$

Definition 1.1.2 *A set* $X \subseteq \Re^n$ *is said to be* convex *if* $\forall$ $\mathbf{x}, \mathbf{y} \in X$ *we have that* X *contains all the convex combinations of* $\mathbf{x}$ *and* $\mathbf{y}$, *i.e.:*

$$\mathbf{z} := [\lambda \mathbf{x} + (1 - \lambda)\mathbf{y}] \in X \ , \ \forall \lambda \in [0, 1].$$

Proposition 1.1.1 *The intersection of two convex sets* $A, B \subseteq \Re^n$ *is still a convex set.*

Proof: Given $\mathbf{x}, \mathbf{y} \in A \cap B$, for all $\lambda \in [0, 1]$ we have $\mathbf{z} := \lambda \mathbf{x} + (1 - \lambda)\mathbf{y} \in A$ by convexity of A, $\mathbf{z} \in B$ by convexity of B, hence $\mathbf{z} \in A \cap B$, as requested. $\square$

Definition 1.1.3 *A function* $f : X \to \Re$ *defined on a convex set* $X \subseteq \Re^n$ *is said to be* convex *if* $\forall$ $\mathbf{x}$, $\mathbf{y}$ $\in X$ *and* $\forall \lambda \in [0, 1]$ *we have that*

$$f(\mathbf{z}) \leq \lambda f(\mathbf{x}) + (1 - \lambda)f(\mathbf{y}) \quad where \quad \mathbf{z} = \lambda \mathbf{x} + (1 - \lambda)\mathbf{y}.$$

There exists a close link between convex sets and convex functions.

Theorem 1.1.1 *Let*

$$X = \{\mathbf{x} \in \Re^n \ : \ g_i(\mathbf{x}) \leq 0 \ , \ i = 1, \ldots, m\}. \tag{1.2}$$

If for all $i \in \{1, \ldots, m\}$ the functions $g_i : \Re^n \to \Re$ are convex, then the set X is convex.

Proof: Clearly

$$X = \bigcap_{i=1}^{m} X_i, \ \text{ where } \ X_i := \{\mathbf{x} \in \Re^n \ : \ g_i(\mathbf{x}) \leq 0\}.$$

By Proposition 1.1.1, it is then sufficient to prove that each set X_i is convex. Indeed, given any two elements $\mathbf{x}$ and $\mathbf{y}$ of X_i and a generic point $\mathbf{z} = \lambda\mathbf{x} + (1 - \lambda)\mathbf{y}$, $\lambda \in [0, 1]$, by the convexity hypothesis of the function g_i we can write

$$g_i(\mathbf{z}) = g_i(\lambda\mathbf{x} + (1 - \lambda)\mathbf{y}) \leq \lambda\, g_i(\mathbf{x}) + (1 - \lambda)g_i(\mathbf{y}) \leq 0,$$

where the latter inequality is valid since $g_i(\mathbf{x}) \leq 0$, $g_i(\mathbf{y}) \leq 0$, and $0 \leq \lambda \leq 1$. It follows that $g_i(\mathbf{z}) \leq 0$, hence $\mathbf{z} \in X_i$. Given the arbitrariness of $\mathbf{x}, \mathbf{y}$ and $\mathbf{z}$, one thus has that X_i is convex, as requested. □

Theorem 1.1.2 *Consider a* convex programming *problem, i.e., a problem* $\min\{f(\mathbf{x}) : \mathbf{x} \in X\}$ *where* $X \subseteq \Re^n$ *is a convex set and* $f : X \to \Re$ *is a convex function. Every locally optimal solution is also a globally optimal solution.*

Proof: Let $\tilde{\mathbf{x}}$ be any locally optimal solution. By the local optimum definition, there exists then $\varepsilon > 0$ such that $f(\tilde{\mathbf{x}}) \leq f(\mathbf{z})$ for all $\mathbf{z} \in I_\varepsilon(\tilde{\mathbf{x}}) := \{\mathbf{x} \in X \ : \ \|\mathbf{x} - \tilde{\mathbf{x}}\| \leq \varepsilon\}$. We have to prove that $f(\tilde{\mathbf{x}}) \leq f(\mathbf{y})$ for all $\mathbf{y} \in X$.

Given any $\mathbf{y} \in X$, consider the point $\mathbf{z}$ belonging to the segment that connects $\tilde{\mathbf{x}}$ to $\mathbf{y}$ and defined as $\mathbf{z} := \lambda\tilde{\mathbf{x}} + (1 - \lambda)\mathbf{y}$, where $\lambda < 1$ is chosen very close to the value 1 so that $\mathbf{z} \in I_\varepsilon(\tilde{\mathbf{x}})$ and hence $f(\tilde{\mathbf{x}}) \leq f(\mathbf{z})$. By the convexity hypothesis of f it follows that

$$f(\tilde{\mathbf{x}}) \leq f(\mathbf{z}) = f(\lambda\tilde{\mathbf{x}} + (1 - \lambda)\mathbf{y}) \leq \lambda f(\tilde{\mathbf{x}}) + (1 - \lambda)f(\mathbf{y}),$$

from which, dividing by $1 - \lambda > 0$, we obtain $f(\tilde{\mathbf{x}}) \leq f(\mathbf{y})$, as requested. □

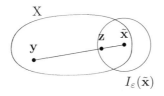

1.2   Linear Programming

A *Linear Programming* (LP), or Linear Optimization, problem is a mathematical programming problem of the kind:

$$\min f(\mathbf{x}) \quad \Rightarrow \quad \min \mathbf{c}^T \mathbf{x}$$

$$g_i(\mathbf{x}) \le 0 \quad \Rightarrow \quad \begin{cases} -\mathbf{a}_i^T \mathbf{x} + b_i \le 0 \quad, \quad i \in \{1, \dots, m\} \\ -x_j \le 0 \quad, \quad j \in \{1, \dots, n\} \end{cases} \tag{1.3}$$

Grouping the row vectors $\mathbf{a}_1^T, \dots, \mathbf{a}_m^T$ into a $m \times n$ matrix A we obtain the compact representation

$$\min\{\mathbf{c}^T \mathbf{x} \,:\, A\mathbf{x} \ge \mathbf{b}\,, \, \mathbf{x} \ge 0\}. \tag{1.4}$$

1.2.1   Example

Consider the following LP problem:

$$\begin{cases} \min & -x_1 & -x_2 & & \\ & 6x_1 & +4x_2 & \le & 24 \\ & 3x_1 & -2x_2 & \le & 6 \\ & x_1 & & \ge & 0 \\ & & x_2 & \ge & 0. \end{cases} \tag{1.5}$$

As shown in Figure 1.3, the set of feasible solutions is convex. The objective function assumes constant values along the isocost lines $-x_1 - x_2 = k$ and reaches its minimum value at point D $= (0, 6)$.   □

1.2.2   Equivalent Forms

An LP problem may be represented in various equivalent ways. In particular, we have the two formulations:

$$\begin{cases} \min \mathbf{c}^T \mathbf{x} \\ A\mathbf{x} \ge \mathbf{b} \\ \mathbf{x} \ge 0 \end{cases} \qquad\qquad \begin{cases} \min \mathbf{c}^T \mathbf{x} \\ A\mathbf{x} = \mathbf{b} \\ \mathbf{x} \ge 0 \end{cases}$$

called *canonical* form (on the left) and *standard* form (on the right), respectively.

The two formulations are equivalent, yet the conversion from one form to the other may implicate the variation of the number of constraints and variables of the problem. The following transformation rules apply:

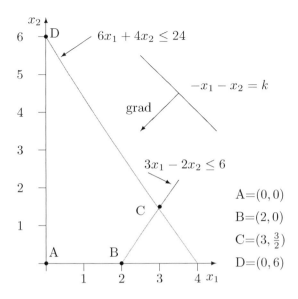

Figure 1.3: Graphical representation.

1. Conversion from "max" to "min":

$$\max \mathbf{w}^T \mathbf{x} = -\min \mathbf{c}^T \mathbf{x} \text{, defining } \mathbf{c}^T = -\mathbf{w}^T$$

2. Constraint conversion from "$\geq$" to "=":

$$\mathbf{a}_i^T \mathbf{x} \geq b_i \Rightarrow \begin{cases} \mathbf{a}_i^T \mathbf{x} - s_i = b_i \\ s_i \geq 0 \end{cases}$$

The new variable s_i is known as *surplus variable* for the i-th constraint.

3. Constraint conversion from "$\leq$" to "=":

$$\mathbf{a}_i^T \mathbf{x} \leq b_i \Rightarrow \begin{cases} \mathbf{a}_i^T \mathbf{x} + s_i = b_i \\ s_i \geq 0 \end{cases}$$

The new variable s_i is known as *slack variable* for the i-th constraint.

4. Free variables:

$$x_i \lessgtr 0 \Rightarrow \begin{cases} x_i = x_i^+ - x_i^- \\ x_i^+ \geq 0 \\ x_i^- \geq 0 \end{cases}$$

Replacing x_i with $x_i^+ - x_i^-$ we eliminate x_i from the problem. Alternatively, x_i may be derived (depending on the other variables) from an equation of the problem in the standard form, and thus replaced everywhere. In this way, we eliminate one equation and one variable.

5. Constraint conversion from "=" to "≥":

$$\mathbf{a}_i^T \mathbf{x} = b_i \quad \Rightarrow \quad \begin{cases} \mathbf{a}_i^T \mathbf{x} & \geq & b_i \\ -\mathbf{a}_i^T \mathbf{x} & \geq & -b_i \end{cases}$$

6. From canonical form to standard form:

$$\min\{\mathbf{c}^T \mathbf{x} \; : \; A\mathbf{x} \geq \mathbf{b} \,, \; \mathbf{x} \geq 0\} \quad \Rightarrow \quad \min\{\mathbf{c}^T \mathbf{x} \; : \; A\mathbf{x} - \mathbf{s} = \mathbf{b} \,, \; \mathbf{x} \geq 0 \,, \; \mathbf{s} \geq 0\}.$$

7. From standard form to canonical form:

$$\min\{\mathbf{c}^T \mathbf{x} \; : \; A\mathbf{x} = \mathbf{b} \,, \; \mathbf{x} \geq 0\} \quad \Rightarrow \quad \min\{\mathbf{c}^T \mathbf{x} \; : \; A\mathbf{x} \geq \mathbf{b} \,, \; -A\mathbf{x} \geq -\mathbf{b} \,, \; \mathbf{x} \geq 0\}.$$

Chapter 2

Linear (Integer) Programming Models

In the following, several simple Linear Programming models are presented, some of which require variables to take on integer values. We talk about *Linear Integer Programming* if all variables must be integer, *Mixed Integer Linear Programming* if only some variables must be integer, and *Binary (or 0-1) Linear Programming* if the variables can only take on the values of 0 and 1.

2.1   Market Survey

An advertising company has to carry out a market survey in order to launch a new product. The survey has to be carried out by telephone, contacting a significant sample of people composed as follows:

type of people	married women	unmarried women	married men	unmarried men
number	≥ 150	≥ 110	≥ 120	≥ 100

Calls can be made in the morning (with an operating cost for the company of 1 Euro/call) and in the evening (with an operating cost of 1.6 Euro/call). The average percentage of people reached is as follows:

Responder	Morning	Evening
married women	30%	30%
unmarried women	10%	20%
married men	10%	30%
unmarried men	10%	15%
nobody	40%	5%

As can be seen, evening calls are more expensive, but allow the company to reach a greater number of people: only 5% of the calls is "in vain". The aim is to minimize the total cost of the calls to be made (morning/evening) so as to reach a significant sample of people.

Choosing as decision variables

- x_1 = number of calls to be made in the morning,

- x_2 = number of calls to be made in the evening,

a possible model is the following:

$$
\begin{array}{lll}
\min & 1x_1 + 1.6x_2 & \text{(total cost of the calls)} \\
& 0.3x_1 + 0.3x_2 \geq 150 & \text{(married women)} \\
& 0.1x_1 + 0.2x_2 \geq 110 & \text{(unmarried women)} \\
& 0.1x_1 + 0.3x_2 \geq 120 & \text{(married men)} \\
& 0.1x_1 + 0.15x_2 \geq 100 & \text{(unmarried men)} \\
& x_1 , x_2 \geq 0 & \text{integers.}
\end{array}
$$

In practice, the integrality constraint may be removed: if the optimal solution is not integer, it will be sufficient to round up the optimal values of x_1 and x_2, with a negligible increase in the overall cost.

2.2   Rental of computers

A company has to rent some computing servers according to the monthly requirements illustrated below:

Month	Jan.	Feb.	Mar.	Apr.	May	Jun.
Quantity	9	5	7	9	10	5

The costs vary depending on the duration of the rental period:

Duration	1 month	2 months	3 months
Cost	400 Euros	700 Euros	900 Euros

We want to decide the rental policy that minimizes the total cost. Let:

- Jan_1, Jan_2, Jan_3 = no. of servers rented in January for 1,2,3 months
- Feb_1, Feb_2, Feb_3 = no. of servers rented in February for 1,2,3 months

  . . .

- Jun_1, Jun_2, Jun_3 = no. of servers rented in June for 1,2,3 months.

A possible model is:

$$
\begin{aligned}
\min \quad & 400(Jan_1 + Feb_1 + Mar_1 + Apr_1 + May_1 + Jun_1) \quad + \\
& 700(Jan_2 + Feb_2 + Mar_2 + Apr_2 + May_2 + Jun_2) \quad + \\
& 900(Jan_3 + Feb_3 + Mar_3 + Apr_3 + May_3 + Jun_3) \\
& Jan_1 + Jan_2 + Jan_3 && \geq 9 && \text{(servers in January)} \\
& Feb_1 + Feb_2 + Feb_3 + Jan_2 + Jan_3 && \geq 5 && \text{(servers in February)} \\
& Mar_1 + Mar_2 + Mar_3 + Feb_2 + Feb_3 + Jan_3 && \geq 7 && \text{(servers in March)} \\
& Apr_1 + Apr_2 + Apr_3 + Mar_2 + Mar_3 + Feb_3 && \geq 9 && \text{(servers in April)} \\
& May_1 + May_2 + May_3 + Apr_2 + Apr_3 + Mar_3 && \geq 10 && \text{(servers in May)} \\
& Jun_1 + Jun_2 + Jun_3 + May_2 + May_3 + Apr_3 && \geq 5 && \text{(servers in June)} \\
& Jan_1, Jan_2, Jan_3, \ldots, Jun_1, Jun_2, Jun_3 && \geq 0 && \text{integers}
\end{aligned}
$$

Note that some variables (May_3, Jun_2, Jun_3) will be set to zero in any optimal solution, and may thus be eliminated from the model.

2.3 Hospital shifts

The aim is to organize the nurses' shifts in a hospital. Each nurse works 5 consecutive days - regardless of the starting day within the week - and is entitled to two days off. Service requirements for the various days of the week require the presence of the following minimum number of nurses:

Day	Monday	Tuesday	Wednesday	Thursday	Friday	Saturday	Sunday
Number	17	13	15	19	14	16	11

We want to organize the service in such a way so as to minimize the total number of nurses.

Let $Mon, Tue, Wed, Thu, Fri, Sat, Sun$ be the number of nurses whose shift begins on Monday, Tuesday, ..., Sunday.

$$
\begin{array}{llllllllll}
\min & Mon & +Tue & +Wed & +Thu & +Fri & +Sat & +Sun \\
& Mon & & & +Thu & +Fri & +Sat & +Sun & \geq & 17 & \text{(attendance on Mon)} \\
& Mon & +Tue & & & +Fri & +Sat & +Sun & \geq & 13 & \text{(attendance on Tue)} \\
& Mon & +Tue & +Wed & & & +Sat & +Sun & \geq & 15 & \text{(attendance on Wed)} \\
& Mon & +Tue & +Wed & +Thu & & & +Sun & \geq & 19 & \text{(attendance on Thu)} \\
& Mon & +Tue & +Wed & +Thu & +Fri & & & \geq & 14 & \text{(attendance on Fri)} \\
& & +Tue & +Wed & +Thu & +Fri & +Sat & & \geq & 16 & \text{(attendance on Sat)} \\
& & & +Wed & +Thu & +Fri & +Sat & +Sun & \geq & 11 & \text{(attendance on Sun)} \\
& Mon, & Tue, & Wed, & Thu, & Fri, & Sat, & Sun & \geq & 0 & \text{integers}
\end{array}
$$

2.4   Service Localization

In a city divided into 6 neighborhoods, there is the need to install a number of Unified Booking Centers (UBC) for health services. A possible installation site has been identified in each neighborhood, and the following average traveling times from each neighborhood to the installation site have been measured (in minutes):

from \ to	1	2	3	4	5	6
1	0*	10*	20	30	30	20
2	10*	0*	25	35	20	10*
3	20	25	0*	15*	30	20
4	30	35	15*	0*	15*	25
5	30	20	30	15*	0*	14*
6	20	10*	20	25	14*	0*

We do not want any user to have a traveling time greater than 15 minutes to get to the nearest UBC and we want to minimize the number of UBCs to be activated.

A possible model has an x_i variable for each site with the following meaning:

$$x_i = \begin{cases} 1 & \text{if the UBC in site } i \text{ is activated} \\ 0 & \text{otherwise} \end{cases}$$

The resulting model is the following:

$$
\begin{array}{llllllll}
\min & x_1 & +x_2 & +x_3 & +x_4 & +x_5 & +x_6 \\
& x_1 & +x_2 & & & & & \geq 1 & \text{(requirements neighb. 1)} \\
& x_1 & +x_2 & & & & +x_6 & \geq 1 & \text{(requirements neighb. 2)} \\
& & & x_3 & +x_4 & & & \geq 1 & \text{(requirements neighb. 3)} \\
& & & x_3 & +x_4 & +x_5 & & \geq 1 & \text{(requirements neighb. 4)} \\
& & & & +x_4 & +x_5 & +x_6 & \geq 1 & \text{(requirements neighb. 5)} \\
& & x_2 & & & +x_5 & +x_6 & \geq 1 & \text{(requirements neighb. 6)} \\
\end{array}
$$

$$0 \leq x_i \leq 1 \text{ integer}, \quad i = 1, \ldots, 6.$$

Note the combinatorial structure of the model that does not directly use the average traveling times but derives the implications in terms of UBCs to be activated.

2.5   Newspapers and TV Advertising

A public transport company has a budget of 150 KEuros (thousands of euros) at its disposal to advertise an initiative through television and print media. A newspaper ad costs 1 KEuro; a maximum of 30 ads of this kind can be made. A TV commercial costs 10 KEuros; at the most, a total of 15 commercials can be made. The number of new users that can be reached with the two kinds of media decreases with the number of ads, with the following rule:

type	slot	new contacts
newspapers	1-10	900
	11-20	600
	21-30	300
television	1-5	10000
	6-10	5000
	11-15	2000

For instance, if we decide to make 12 newspaper ads and 8 television commercials we will reach 75.200 new users (9000 + 1200 via newspapers; 50.000 + 15.000 via television), with a total cost of 12 + 80 = 92 KEuro. Note that the number of contacts decreases with the various slots (saturation effect).

We want to maximize the number of contacts, still fitting into the total budget of 150 KEuros.

Let g_1, g_2, g_3 and t_1, t_2, t_3 be the number of ads to be made on newspapers and in television in the various slots. A possible model is:

$$
\begin{aligned}
\max \quad & 900g_1 + 600g_2 + 300g_3 + 10000t_1 + 5000t_2 + 2000t_3 \\
& g_1 + g_2 + g_3 + 10t_1 + 10t_2 + 10t_3 \leq 150 \qquad \text{(budget)} \\
& 0 \leq g_i \leq 10 \ \text{integers}, \ i = 1, 2, 3 \qquad \text{(newspapers' slots)} \\
& 0 \leq t_i \leq 5 \ \text{integers}, \ i = 1, 2, 3 \qquad \text{(TV's slots)}
\end{aligned}
$$

Actually, the model is not complete, as there are no constraints that force the activation of a slot only when the previous ones have been saturated. For instance, the model considers the solution $g_1 = 8$, $g_2 = 3$, $g_3 = 0$, $t_1 = t_2 = t_3 = 0$ as feasible. However, such an inconsistency cannot occur in an optimal solution, as the alternative solution $g_1 = 10$, $g_2 = 1$, $g_3, t_1, t_2, t_3 = 0$ has the same cost but allows us to reach a greater number of contacts. This ensures that the optimal solution will use the various slots in a consistent manner.

2.6   Blending of products

A refinery produces three types of gasoline (A, B, C), each of which is obtained by mixing 4 basic products. The availability and unit price of the basic products, and the composition of the various kinds of gasoline and the corresponding unit revenue figures are shown below. The aim is to maximize the total net profit (revenue minus cost).

Product	Availability	Cost
1	3000	3
2	2000	6
3	4000	4
4	1000	5

gasoline \ product	1	2	3	4	revenue
A	$\leq 30\%$	$\geq 40\%$	$\leq 50\%$	$\times$	5.5
B	$\leq 50\%$	$\geq 10\%$	$\times$	$\times$	4.5
C	$\geq 70\%$	$\times$	$\times$	$\times$	3.5

For instance, gasoline B requires no more than 50% of product 1, at least 10% of product 2, and can use products 3 and 4 without restrictions.

To formulate the model, we choose as variables the quantities y_A, y_B, y_C of gasoline A, B and C produced, and the quantities x_{ij} that indicate the quantity of product $i \in \{1, 2, 3, 4\}$ that is used in the production of gasoline $j \in \{A, B, C\}$. The objective function to maximize is obtained calculating:

$$\underbrace{5.5y_A + 4.5y_B + 3.5y_C}_{\text{gasoline revenue}} - \underbrace{3(x_{1A} + x_{1B} + x_{1C})}_{\text{product cost 1}} - \underbrace{6(x_{2A} + x_{2B} + x_{2C})}_{\text{product cost 2}}$$

$$- \underbrace{4(x_{3A} + x_{3B} + x_{3C})}_{\text{product cost 3}} - \underbrace{5(x_{4A} + x_{4B} + x_{4C})}_{\text{product cost 4}}.$$

We obtain:

$$\max \quad 5.5y_A + 4.5y_B + 3.5y_C$$
$$-3x_{1A} - 3x_{1B} - 3x_{1C} - 6x_{2A} - 6x_{2B} - 6x_{2C}$$
$$-4x_{3A} - 4x_{3B} - 4x_{3C} - 5x_{4A} - 5x_{4B} - 5x_{4C}$$

$$\left.\begin{array}{l} x_{1A} + x_{1B} + x_{1C} \leq 3000 \\ x_{2A} + x_{2B} + x_{2C} \leq 2000 \\ x_{3A} + x_{3B} + x_{3C} \leq 4000 \\ x_{4A} + x_{4B} + x_{4C} \leq 1000 \end{array}\right\} \quad \text{(availability constraints)}$$

$$\left.\begin{array}{l} y_A = x_{1A} + x_{2A} + x_{3A} + x_{4A} \\ y_B = x_{1B} + x_{2B} + x_{3B} + x_{4B} \\ y_C = x_{1C} + x_{2C} + x_{3C} + x_{4C} \end{array}\right\} \quad \text{(consistency constraints)}$$

$$\left.\begin{array}{l} x_{1A} \leq 0.3y_A \\ x_{2A} \geq 0.4y_A \\ x_{3A} \leq 0.5y_A \\ x_{1B} \leq 0.5y_B \\ x_{2B} \geq 0.1y_B \\ x_{1C} \geq 0.7y_C \end{array}\right\} \quad \text{(gasoline restrictions)}$$

$$y_A, y_B, y_C, x_{1A}, \ldots, x_{4C} \geq 0$$

Since in this example $\mathbf{x} \geq 0$ implies $\mathbf{y} \geq 0$, the variables y_A, y_B, y_C may be eliminated from the model, deriving them from the three equations and replacing them everywhere.

2.7   Assignment of machining operations

A company can produce a carburetor model through three different production lines (P_1, P_2, P_3), using three machines (M_A, M_B, M_C). Each production line uses semi-finished parts, different for each line, which must be assembled on the three machines (shared among the lines) according to the following specifications:

production chain	P_1	P_2	P_3
unit cost of the parts (Euro)	6	3	7
minutes on M_A	3	2	6
minutes on M_B	4	8	2
minutes on M_C	6	3	1

machine	cost (Euro/min)	availability (min)
M_A	5	2000
M_B	4	3000
M_C	3	600

For instance, the production of a carburetor through line P_1 costs 6 Euros for the supply of the semi-finished parts, and employs machines M_A, M_B and M_C for 3, 4 and 6 minutes, respectively; machine M_A costs 5 Euros per minute and is available for 2000 minutes.

The unit sales price of the end product is 90 Euros. We want to maximize the overall profit.

Let then x_i be the number of carburetors produced by line P_i ($i = 1, 2, 3$). The overall profit can be calculated as:

$$\underbrace{90(x_1 + x_2 + x_3)}_{\text{revenue}} - \underbrace{(6x_1 + 3x_2 + 7x_3)}_{\text{cost of the parts}} - \underbrace{5(3x_1 + 2x_2 + 6x_3)}_{\text{cost on } M_A}$$

$$- \underbrace{4(4x_1 + 8x_2 + 2x_3)}_{\text{cost on } M_B} - \underbrace{3(6x_1 + 3x_2 + 1x_3)}_{\text{cost on } M_C} = 35x_1 + 36x_2 + 42x_3.$$

Hence, the model reads:

$$
\begin{array}{rlllll}
\max & 35x_1 & +36x_2 & +42x_3 \\
& 3x_1 & +2x_2 & +6x_3 & \leq & 2000 & (M_A \text{ availability}) \\
& 4x_1 & +8x_2 & +2x_3 & \leq & 3000 & (M_B \text{ availability}) \\
& 6x_1 & +3x_2 & +x_3 & \leq & 600 & (M_C \text{ availability}) \\
& x_1, & x_2, & x_3 & \geq & 0 & \text{integers}
\end{array}
$$

2.8 Production of sausages

A sausage factory produces three kinds of sausage (P_1, P_2, P_3), which can be sold as they are or can be smoked. In normal operation, the smokehouse can process 420 sausages, and in extraordinary operation it can process 250 additional parts. The profits of the various kinds of sausage (in Euro), depending on the treatment, are reported in the table below. We want to maximize the total sales profit.

sausage	daily production	profit normal op.	profit ord. op. smoked	profit extra op. smoked
P_1	480	8	14	11
P_2	400	4	12	7
P_3	230	4	13	9

For instance, the first row of the table should be interpreted as follows. 480 pieces of sausage P_1 are produced daily. Each non-smoked piece yields a profit of 8 Euros, and of 14 Euros if smoked (but only 11 Euros if considering the cost for the extra operation).

For the formulation of the model, the production levels for each sausage are chosen as variables, for each of the possible treatments. We have thus variables x_{ij}, where $i \in \{1, 2, 3\}$ indicates the kind of sausage and $j \in \{N, O, E\}$ indicates the treatment (**N**ot smoked, **O**rdinary smoked, **E**xtra smoked). Thus, the corresponding model is:

$$
\begin{array}{ll}
\max & 8x_{1N} + 14x_{1O} + 11x_{1E} + 4x_{2N} + 12x_{2O} + 7x_{2E} + 4x_{3N} + 13x_{3O} + 9x_{3E} \\
& x_{1N} + x_{1O} + x_{1E} = 480 \quad (P_1 \text{ production}) \\
& x_{2N} + x_{2O} + x_{2E} = 400 \quad (P_2 \text{ production}) \\
& x_{3N} + x_{3O} + x_{3E} = 230 \quad (P_3 \text{ production}) \\
& x_{1O} + x_{2O} + x_{3O} \leq 420 \quad (\text{ord. capacity}) \\
& x_{1E} + x_{2E} + x_{3E} \leq 250 \quad (\text{extra capacity}) \\
& x_{1N}, \ldots, x_{3E} \geq 0 \quad \text{integers}
\end{array}
$$

2.9   Radio Production

An electronic company needs to produce at least 20,000 radios in a 4-week period. The unit revenues from the sale of the radios each week are as follows:

Week	1	2	3	4
Revenue (Euro)	20	18	16	14

The company initially has 40 workers, each producing 50 radios per week. It is also possible to hire apprentice workers, three of whom can be trained by a worker within a 1-week period. During the training period the teacher worker does not produce any radio. A worker costs 200 Euros per week, while an apprentice costs 100 Euros for the apprenticeship week. Each radio needs basic components for a cost of 2 Euros. We want to maximize the overall profit.

For each $i \in \{1, \ldots, 4\}$, possible decision variables are:

- x_i = number of workers in charge of production during week i

- y_i = number of workers in charge of apprenticeship during week i

- z_i = number of apprentices trained during week i

Simple considerations allow us to eliminate from the model the variables y_4 and z_4, as during the last week it is not convenient to train apprentices. The overall profit can be expressed as:

$$\underbrace{50(20x_1 + 18x_2 + 16x_3 + 14x_4)}_{\text{radio revenues}} -$$

$$\underbrace{200(4 \cdot 40 + 3z_1 + 2z_2 + z_3)}_{\text{wages of workers}} - \underbrace{100(z_1 + z_2 + z_3)}_{\text{wages of apprentices}} - \underbrace{2(50x_1 + 50x_2 + 50x_3 + 50x_4)}_{\text{cost of components}}$$

Among the workers we consider both those on duty from the first week, and those who become workers after the apprenticeship.

Hence, the model is:

$$-32000 + \max \quad 900x_1 + 800x_2 + 700x_3 + 600x_4 - 700z_1 - 500z_2 - 300z_3$$

$$50(x_1 + x_2 + x_3 + x_4) \geq 20000 \quad \text{(minimum production constraint)}$$

$$\left. \begin{aligned} z_1 &\leq 3y_1 \\ z_2 &\leq 3y_2 \\ z_3 &\leq 3y_3 \end{aligned} \right\} \quad \text{(trainable apprentices)}$$

$$\left. \begin{aligned} x_1 + y_1 &\leq 40 \\ x_2 + y_2 &\leq 40 + z_1 \\ x_3 + y_3 &\leq 40 + z_1 + z_2 \\ x_4 &\leq 40 + z_1 + z_2 + z_3 \end{aligned} \right\} \quad \text{(availability of workers)}$$

$$x_1, \ldots, z_3 \geq 0 \quad \text{integers}$$

2.10 Scheduling of activities

A company produces sofas, each of which needs a series of processing steps linked by precedence relations. Each processing has a weight and a duration. The first processing step conventionally begins at $t = 0$. Our aim is to determine the start times of each processing step so as to minimize the weighted average. The problem data is summarized in the following table.

Processing Step	Weight	Duration	Previous Steps
1	0	10	—
2	2	23	1
3	5	12	1
4	7	10	2
5	10	3	3, 4
6	12	11	2
7	17	5	2
8	21	4	7
9	26	10	5, 6

For instance, processing step 5 cannot begin before the end of steps 3 and 4, has a weight equal to 10, and a duration equal to 3.

Indicating with t_i the beginning of processing step $i \in \{1, \ldots, 9\}$, we obtain the model:

$$\tfrac{1}{100} \min \quad 2t_2 + 5t_3 + 7t_4 + 10t_5 + 12t_6 + 17t_7 + 21t_8 + 26t_9$$

$$t_1 = 0$$
$$t_2 \geq t_1 + 10$$
$$t_3 \geq t_1 + 10$$
$$t_4 \geq t_2 + 23$$
$$t_5 \geq t_3 + 12$$
$$t_5 \geq t_4 + 10$$
$$t_6 \geq t_2 + 23$$
$$t_7 \geq t_2 + 23$$
$$t_8 \geq t_7 + 5$$
$$t_9 \geq t_5 + 3$$
$$t_9 \geq t_6 + 11$$
$$t_1, \ldots, t_9 \geq 0$$

2.11 Transport of Containers

A transport company has to transfer empty containers from its warehouses to the main national ports. The availability of empty containers at the warehouses and the requests at the ports are as follows:

availability at warehouses

Verona	10
Perugia	12
Rome	20
Pescara	24
Taranto	18
Lamezia Terme	40

requests at ports

Genua	20
Venice	15
Ancona	25
Naples	33
Bari	21

To a first approximation, transport costs can be considered proportional to the number of containers and to the kilometers traveled by the trucks: cost :=[no. of transported containers]×[Km travelled]. We want to determine the minimum-cost transportation policy.

The writing of the corresponding model is left as an exercise for the reader (using a ruler, distances can be measured in millimeters between the various locations on a map).

Chapter 3

Advanced Models

We will now briefly describe some models of (Mixed) Integer Linear Programming of considerable interest in terms of application. Some of these models will be further analyzed in Chapter 8.

3.1   Big-M constraints

Let us start by describing some of the modeling "tricks" widely used in practice, which enable us to easily model apparently nonlinear situations—even though the resolution of the resulting models can be remarkably burdensome from the computational point of view.

Conditional Deactivation of a Constraint

In some cases, there is the need to model the possibility that a constraint of the model

$$\mathbf{a}_i^T \mathbf{x} \geq b_i \tag{3.1}$$

will be deactivated under certain conditions. If the conditions are known or easily computable starting from the problem input data, we can simply omit to write the constraint. If this is not the case, it is possible to introduce a binary variable $y_i \in \{0, 1\}$ expressing the will to deactivate the constraint ($y_i = 1$) or to leave it active ($y_i = 0$), imposing the logical implication

$$y_i = 0 \Rightarrow \mathbf{a}_i^T \mathbf{x} \geq b_i$$

through constraints

$$\mathbf{a}_i^T \mathbf{x} \geq b_i - M \, y_i \tag{3.2}$$

$$y_i \in \{0, 1\} \tag{3.3}$$

Constant $M \gg 0$ that appears in constraint (3.2) is a very large value that serves to deactivate the constraint if the corresponding variable y is equal to 1, and is commonly called "Big-M". Of course, variable y_i will then appear in the objective function and/or in other constraints of the problem, in order to correctly model any interactions with other constraints and/or costs corresponding to the deactivation of the constraint in question.

Note that the determination of a "sufficiently big" value M in order to deactivate the constraint requires an overestimation of the maximum value that quantity $b_i - \mathbf{a}_i^T \mathbf{x}$ can take in a feasible solution of the problem—if such quantity were unbounded from above, the method could not be applied. In practice, we are often satisfied with fairly coarse estimates, avoiding the use of a value M unnecessarily big especially to limit the numerical problems linked to the presence of coefficients with a very different order of magnitude in the same constraint.

Disjunctive Constraints

Let us suppose to have the following two constraints

$$\mathbf{a}_i^T \mathbf{x} \geq b_i \tag{3.4}$$
$$\mathbf{a}_k^T \mathbf{x} \geq b_k \tag{3.5}$$

and to impose that *at least* one of them is verified (i.e., it is not necessary to satisfy both of them). Introducing for each constraint a binary variable to deactivate the corresponding constraint, we obtain

$$\mathbf{a}_i^T \mathbf{x} \geq b_i - M_i \, y_i \tag{3.6}$$
$$\mathbf{a}_k^T \mathbf{x} \geq b_k - M_k \, y_k \tag{3.7}$$
$$y_i + y_k \leq 1 \tag{3.8}$$
$$y_i, \, y_k \in \{0,1\} \tag{3.9}$$

where constants $M_i, M_k \gg 0$ that appear in constraints (3.6) and (3.7) act as Big-M's. Constraint (3.8) establishes that at most one of the variables y_i and y_k may be equal to 1, thus preventing both constraints from being deactivated.

Functions with a Discrete Set of Values

Let us suppose we are in the situation where a linear function $\mathbf{f}^T \mathbf{x}$ (the objective function or the left hand side of a constraint) can take only values belonging to a discrete set, i.e., we want to to impose

$$\mathbf{f}^T \mathbf{x} \in \{v_1, \dots, v_k\}$$

where $v_1, \dots, v_k$ are input constants (and not variables). Introducing a binary variable y_i for each possible value v_i, we can write a model of the type

$$\mathbf{f}^T \mathbf{x} = \sum_{i=1}^{k} v_i \, y_i$$

$$\sum_{i=1}^{k} y_i = 1$$

$$y_i \in \{0,1\}, \; \forall i = 1, \ldots, k$$

where the generic variable y_i is equal to 1 if the function is equal to v_i, otherwise it is equal to 0.

Fixed Charge

Let us suppose that we now have to describe a minimization model in which the objective function includes fixed costs. This situation arises, for example, in production systems in which the production of a product j has a unit cost c_j, but to produce type j products (in any quantity) there is a *one-off* fixed cost $k_j > 0$ (purchase of machinery, training of personnel, etc.). In these cases, the contribution to the objective function deriving by type j products can be expressed as

$$f_j(x_j) = \begin{cases} k_j + c_j\,x_j & \text{if } x_j > 0 \\ 0 & \text{if } x_j = 0 \end{cases} \tag{3.10}$$

where variable x_j represents the produced quantity of type j products. Obviously, the function $f_j(x_j)$ described by (3.10) is nonlinear since it presents a discontinuity for $x_j = 0$.

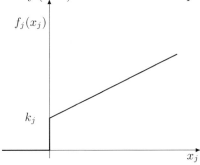

Figure 3.1: Function with fixed charge.

To model this situation, it is possible to introduce a binary variable y_j that is equal to 1 if $x_j > 0$, describing function $f_j(x_j)$ as

$$f_j(x_j) = k_j\,y_j + c_j\,x_j \tag{3.11}$$

$$x_j \le M\,y_j \tag{3.12}$$

$$y_j \in \{0,1\} \tag{3.13}$$

where $M \gg 0$ is the usual Big-M. Constraint (3.12) forces variable y_j to be equal to 1 if $x_j > 0$, thus paying the fixed cost k_j. Vice versa, if $x_j = 0$ then constraint (3.12) becomes $0 \le M\,y_j$ and hence it is deactivated, allowing variable y_j to be equal to zero—hence the objective function (to be minimized) does not have to pay the (positive) fixed charge k_j.

Binary Representation of Integer Variables

In some cases, it can be useful to convert an Integer Linear Programming model with generic integer variables in a Binary Linear Programming model in which all integer variables are binary.

Suppose we have a problem in which variable x_j may take on all integer values from 0 to an assigned integer $u_j \geq 2$. This variable can be represented by means of a set of $k+1$ binary variables in the following way

$$x_j = \sum_{i=0}^{k} 2^i y_i^j \tag{3.14}$$

$$x_j \leq u_j \tag{3.15}$$

$$y_i^j \in \{0,1\}, \ \forall i = 0, \ldots, k \tag{3.16}$$

where k is the smallest integer such that $u_j < 2^{k+1}$. Replacing x_j with its expression (3.14) in the objective function and in all constraints of the model, we obtained the requested Binary Linear Programming model.

3.2   Knapsack Problem 0-1

Give a set $N = \{1, \ldots, n\}$ of objects, the j-th with profit $p_j > 0$ and weight $w_j > 0$, and a container (*knapsack*) with capacity W, we want to select a set of objects to be inserted in the container so that:

- the total weight of the selected objects does not exceed W;

- the profit of the selected objects is a maximum.

Introducing the binary variables

$$x_j = \begin{cases} 1 & \text{if object } j \text{ is selected} \\ 0 & \text{otherwise} \end{cases} \qquad (j \in N)$$

we obtain the model

$$\max \sum_{j \in N} p_j\, x_j \tag{3.17}$$

$$\sum_{j \in N} w_j\, x_j \leq W \tag{3.18}$$

$$x_j \in \{0,1\} \qquad j \in N \tag{3.19}$$

Change-Making Problem

Given a set $N = \{1, \ldots, n\}$ of objects, the j-th of which has a value of $p_j > 0$ and is available in k_j samples, and a reference value W, we want to select a number of samples for each object so that

- the total value of the selected objects is exactly equal to W,

- the total number of samples selected is minimal.

An application is the change-making problem: each object j corresponds to a type of coin (e.g. 10, 20, 50 Euro cents, and 1, 5, 10, 20, 50, 100 Euros), there is a limited number of samples k_j in the cash register, and we want to reach a certain value W (the change to be returned to the customer) minimizing the number of coins to be used.

Model:

$$\min \sum_{j \in N} x_j \tag{3.20}$$

$$\sum_{j \in N} p_j x_j = W \tag{3.21}$$

$$0 \le x_j \le k_j, \ x_j \text{ integer}, \qquad j \in N \tag{3.22}$$

Multiple Knapsack

This is a generalization of the Knapsack Problem in which a set $M = \{1, \ldots, m\}$ of containers is given, the i-th of which has capacity W_i. The problem is well defined if each object can be inserted in at least one container, and if each container can contain at least one object, i.e., if

$$\max_{j \in N} w_j \le \max_{i \in M} W_i \quad \text{and} \quad \min_{i \in M} W_i \ge \min_{j \in N} w_j$$

A mathematical model is based on the following binary variables

$$x_{ij} = \begin{cases} 1 & \text{if object } j \text{ is inserted in container } i \\ 0 & \text{otherwise} \end{cases} \qquad (i \in M, \ j \in N)$$

and reads

$$\max \sum_{j \in N} p_j \sum_{i \in M} x_{ij} \tag{3.23}$$

$$\sum_{j \in N} w_j \, x_{ij} \leq W_i \qquad i \in M \tag{3.24}$$

$$\sum_{i \in M} x_{ij} \leq 1 \qquad j \in N \tag{3.25}$$

$$x_{ij} \in \{0, 1\} \qquad i \in M, \; j \in N \tag{3.26}$$

3.3　Job scheduling

Given a set $N = \{1, \dots, n\}$ of jobs, the j-th of which requires a processing time p_j, and a set $M = \{1, \dots, m\}$ of identical machines, we want to assign the jobs to the machines so that

- each job is assigned exactly to one machine;

- each machine processes at most one job at a time;

- the makespan (i.e., the maximum uptime of all the machines) is minimal.

Note that the processing order within the same machine is not relevant.

Example $n = 5$, $p = (100, 80, 70, 40, 30)$, $m = 2$

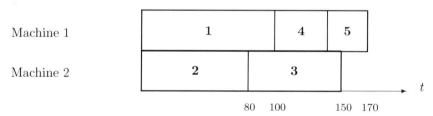

The drawn solution has value (makespan) equal to $\max\{150, 170\} = 170$.

A possible model of the problem is the following:

$$x_{ij} = \begin{cases} 1 & \text{if job } j \text{ is assigned to machine } i \\ 0 & \text{otherwise} \end{cases} \qquad (i \in M, \; j \in N)$$

$$z = \text{completion time of the last job (makespan)}$$

$$\min \ z \tag{3.27}$$

$$z \geq \sum_{j \in N} p_j \, x_{ij} \qquad i \in M \tag{3.28}$$

$$\sum_{i \in M} x_{ij} = 1 \qquad j \in N \tag{3.29}$$

$$x_{ij} \in \{0, 1\} \qquad i \in M, \ j \in N \tag{3.30}$$

3.4   Assignment Problem

Given a square matrix C with n rows and n columns, whose generic element c_{ij} represents the cost to assign row i to column j, the Assignment Problem consists in assigning the rows to the columns so that:

- each row is assigned to one and only one column;

- each column is assigned one and only one row;

- the total cost of the assignment is minimal.

Example $n = 5$ $\quad C = \begin{bmatrix} 4 & 6 & 9 & \underline{3} & 10 \\ \underline{2} & 4 & 11 & 8 & 5 \\ 5 & 7 & \underline{5} & 16 & 9 \\ 6 & 8 & 12 & 4 & \underline{7} \\ 13 & \underline{10} & 8 & 7 & 3 \end{bmatrix}$

The underlined elements identify a (non-optimal) solution of value $3+2+5+7+10=27$.

The problem can be written as:

$$x_{ij} = \begin{cases} 1 & \text{if row } i \text{ is assigned to column } j \\ 0 & \text{otherwise} \end{cases} \qquad (i, j = 1, \ldots, n)$$

$$\min \sum_{i=1}^{n} \sum_{j=1}^{n} c_{ij} \, x_{ij} \qquad\qquad (3.31)$$

$$\sum_{j=1}^{n} x_{ij} = 1 \qquad i = 1, \ldots, n \qquad\qquad (3.32)$$

$$\sum_{i=1}^{n} x_{ij} = 1 \qquad j = 1, \ldots, n \qquad\qquad (3.33)$$

$$x_{ij} \in \{0, 1\} \qquad i, j = 1, \ldots, n \qquad\qquad (3.34)$$

It can be proved that the integrality condition in (3.34) is redundant and can therefore be eliminated.

3.5   Set Covering Problem

Consider a binary matrix A with m rows and n columns, the j-th of which is associated with a cost c_j. Given a row i and a column j, if $a_{ij} = 1$ we say that column j *covers* row i. The Set Covering Problem calls for a minimum-cost subset of the columns such that all rows are covered. In other words, we want to select a subset of columns S such that:

  - for each row i there exists at least one column $j \in S$ with $a_{ij} = 1$;

  - the overall cost of the columns in S is minimal.

An obvious necessary and sufficient condition for the problem to be solved is that each row i is covered by at least one column j, i.e., for each $i = 1, \ldots, m$, there exists at least a $j \in \{1, \ldots, n\}$ such that $a_{ij} = 1$. When the problem admits a solution, finding a feasible solution is trivial (we just have to choose $S = \{1, \ldots, n\}$), while finding the optimal solution can be extremely difficult.

Example $n = 10$, $m = 6$,

$$c = \begin{bmatrix} 15 & 20 & 14 & 22 & 28 & 16 & 30 & 18 & 19 & 21 \end{bmatrix}$$

$$A = \begin{bmatrix} 1 & 1 & 0 & 0 & 0 & 0 & 0 & 0 & 1 & 0 \\ 0 & 1 & 0 & 1 & 1 & 1 & 0 & 1 & 0 & 1 \\ 1 & 0 & 1 & 1 & 0 & 1 & 1 & 0 & 1 & 0 \\ 0 & 1 & 1 & 0 & 1 & 0 & 0 & 0 & 1 & 1 \\ 0 & 0 & 0 & 1 & 0 & 0 & 1 & 1 & 0 & 0 \\ 1 & 0 & 1 & 1 & 1 & 0 & 0 & 0 & 0 & 1 \end{bmatrix}$$

A feasible solution with cost $22 + 19$ is given by $S = \{4, 9\}$.

A model of the problem is

$$x_j = \begin{cases} 1 & \text{if column } j \text{ is selected in } S \\ 0 & \text{otherwise} \end{cases} \quad (j = 1, \ldots, n)$$

$$\min \sum_{j=1}^{n} c_j \, x_j \tag{3.35}$$

$$\sum_{j=1}^{n} a_{ij} \, x_j \geq 1 \qquad i = 1, \ldots, m \tag{3.36}$$

$$x_j \in \{0, 1\} \qquad j = 1, \ldots, n \tag{3.37}$$

3.6 Set Partitioning Problem

This is a variant of the Set Covering Problem in which each row must be covered *exactly* once. The model is similar to that of the Set Covering but constraints (3.36) must be rewritten with the equality sign, which, in practice, makes the problem even more difficult to solve.

The necessary condition for the problem to admit a solution is that for each row i there exists at least a column j such that $a_{ij} = 1$. However, no simple (i.e., efficiently verifiable) sufficient conditions are known to guarantee the existence of a feasible solution. In other words, even just deciding whether the problem admits a solution can be extremely difficult.

3.7   Facility (or Plant) Location Problem

Given

- a set M of customers to be served, the i-th of which has a demand $d_i > 0$;

- a set N of possible facilities to be used, the j-th of which has an activation cost $f_j > 0$ and a maximum productive capacity $b_j > 0$;

- the transportation cost $c_{ij} \geq 0$ for unit of product from plant $j \in N$ to customer $i \in M$

decide which facilities to build and which facility is to be used for serving each customer so that

- each customer is completely served by one and only one facility;

- each facility, if activated, serves a subset of customer whose overall demand does not exceed the productive capacity of the facility itself;

- the overall cost (activation+transportation) is minimized.

Model:

$$y_j = \begin{cases} 1 & \text{if facility } j \text{ is activated} \\ 0 & \text{otherwise} \end{cases} \qquad (j \in N)$$

$$x_{ij} = \begin{cases} 1 & \text{if customer } i \text{ is served by facility } j \\ 0 & \text{otherwise} \end{cases} \qquad (i \in M,\ j \in N)$$

$$\min \sum_{j \in N} f_j\, y_j + \sum_{j \in N} \sum_{i \in M} c_{ij}\, (d_i\, x_{ij}) \tag{3.38}$$

$$\sum_{j \in N} x_{ij} = 1 \qquad i \in M \tag{3.39}$$

$$\sum_{i \in M} d_i\, x_{ij} \leq b_j\, y_j \qquad j \in N \tag{3.40}$$

$$x_{ij} \in \{0,1\} \qquad i \in M,\ j \in N \tag{3.41}$$

$$y_j \in \{0,1\} \qquad j \in N \tag{3.42}$$

There exists a variant of the problem in which each customer is allowed to be served by different facilities. In this case, we have a similar model, with

$$x_{ij} = \text{fraction of the demand of customer } i \text{ served by facility } j$$

and in which constraints (3.41) are replaced by

$$0 \le x_{ij} \le 1 \qquad i \in M, \ j \in N \tag{3.43}$$

A particular case of the problem is known as the *Uncapacitated* Facility Location Problem: the demands of all customer are equal to each other (e.g, $d_i = 1$ for all i) and the capacities of the facilities are infinite ($b_j = M \gg 0$ for all j). In this case, it is possible to write a more efficient model than (3.38)–(3.42), replacing constraints (3.40), rewritten as

$$\sum_{i \in M} x_{ij} \le M \ y_j \qquad j \in N \tag{3.44}$$

with the set of (tighter) constraints:

$$x_{ij} \le y_j \qquad i \in M, \ j \in N \tag{3.45}$$

3.8   Vertex Cover Problem

Given an undirected graph (see Chapter 7) $G = (V, E)$ and a cost $c_i > 0$ associated with each vertex $i \in V$, determine a subset of vertices such that:

- for each edge, at least one of the associated nodes is selected;

- the sum of the costs of the selected vertices is minimal.

In other words, a selected vertex covers all edges incident to it, and the problem requires to covers all edges of the graph using a set of vertices with overall minimal cost.

Model:

$$x_i = \begin{cases} 1 & \text{if vertex } i \text{ is selected} \\ 0 & \text{otherwise} \end{cases} \qquad (i \in V)$$

$$\min \sum_{i \in V} c_i \ x_i \tag{3.46}$$

$$x_i + x_j \ge 1 \qquad [i, j] \in E \tag{3.47}$$

$$x_i \in \{0, 1\} \qquad i \in V \tag{3.48}$$

3.9   Matching Problem

Given an undirected graph $G = (V, E)$ with costs c_e on the edges and $n = |V|$ even, select $n/2$ edges so that:

- for each vertex $v \in V$, an edge incident to v is selected;

- the overall cost of the selected edges is minimal.

In practice, we have to match the vertices of the graph two by two by means of a set of edges with minimal cost.

Model:

$$
x_e = \begin{cases} 1 & \text{if edge } e \text{ is selected} \\ 0 & \text{otherwise} \end{cases} \qquad (e \in E)
$$

$$
\min \sum_{e \in E} c_e \, x_e \tag{3.49}
$$

$$
\sum_{e \in \delta(v)} x_e = 1 \qquad v \in V \tag{3.50}
$$

$$
x_e \in \{0, 1\} \qquad e \in E \tag{3.51}
$$

where $\delta(v)$ is the set of incident edges in vertex v. One can prove that, if G is bipartite (i.e., if it does not contain any odd cycle), then the integrality condition in (3.51) is redundant.

Chapter 4

Simplex Algorithm

Some fundamental geometric and algorithmic aspects of linear programming will now be considered.

4.1 Geometry of Linear Programming

The set of feasible solutions of an LP problem is defined by linear inequalities, each of which identifies a region of the vector space $\Re^n$.

Definition 4.1.1 *The sets* $\{\mathbf{x} \in \Re^n \; : \; \alpha^T \mathbf{x} \le \alpha_0\}$ *and* $\{\mathbf{x} \in \Re^n \; : \; \alpha^T \mathbf{x} = \alpha_0\}$ *are called* affine half-space *and* hyperplane, *respectively, induced by* (α, α_0).

Definition 4.1.2 *A (convex) polyhedron is defined as the intersection of a finite number of affine half-spaces and hyperplanes.*

The sets of feasible solutions of LP problems are hence polyhedra.

Definition 4.1.3 *A bounded (i.e., there exists $M > 0$ such that $||\mathbf{x}|| \le M$ for all $\mathbf{x} \in P$) polyhedron P is called* polytope.

Definition 4.1.4 *A point $\mathbf{x}$ of a polyhedron P is said to be an* extreme point *or a* vertex *of P if it cannot be expressed as a strict convex combination of other two points of the polyhedron, i.e., if there exist no $\mathbf{y}, \mathbf{z} \in P$, $\mathbf{y} \ne \mathbf{z}$ and $\lambda \in (0,1)$ such that $\mathbf{x} = \lambda \mathbf{y} + (1 - \lambda)\mathbf{z}$.*

Each polyhedron has a finite number of vertices. In the case of bounded polyhedra, it is also possible to demonstrate the following fundamental result, known as the *Minkowski-Weyl Theorem*:

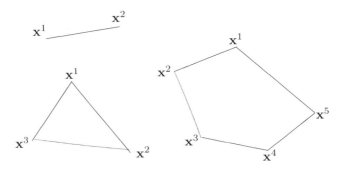

Figure 4.1: Polytopes and Vertices

Theorem 4.1.1 *Every point of a polytope can be obtained as the convex combination of its vertices.*

An important consequence of the theorem is the following:

Theorem 4.1.2 *If the set P of the feasible solutions of the linear programming problem $\min\{\mathbf{c}^T\mathbf{x} : \mathbf{x} \in P\}$ is bounded, then there exists at least one optimal vertex of P.*

Proof: Let $\mathbf{x}^1, \ldots, \mathbf{x}^k$ be the vertices of P and $z^* := \min\{\mathbf{c}^T\mathbf{x}^i : i = 1, \ldots, k\}$. Given any $\mathbf{y} \in P$, we need to prove that $\mathbf{c}^T\mathbf{y} \geq z^*$. Indeed, $\mathbf{y} \in P$ implies the existence of multipliers $\lambda_1, \ldots, \lambda_k \geq 0$, $\sum_{i=1}^k \lambda_i = 1$, such that $\mathbf{y} = \sum_{i=1}^k \lambda_i \mathbf{x}^i$. Hence we have

$$\mathbf{c}^T\mathbf{y} = \mathbf{c}^T \sum_{i=1}^k \lambda_i \mathbf{x}^i = \sum_{i=1}^k \lambda_i(\mathbf{c}^T\mathbf{x}^i) \geq \sum_{i=1}^k \lambda_i z^* = z^*.$$

$\square$

4.1.1 Vertices and basic solutions

The fact that an optimal solution of a bounded LP problem coincides with a vertex suggests the following resolution procedure: we start from any vertex and then iteratively move onto a better "adjacent" vertex, until we reach a locally, and thus globally, optimal vertex. The procedure requires an algebraic characterization of the vertices of P.

Consider example (1.5), rewritten in standard form:

$$\begin{cases} \min & -x_1 & -x_2 \\ & 6x_1 & +4x_2 & +x_3 & & = & 24 \\ & 3x_1 & -2x_2 & & +x_4 & = & 6 \\ & x_1, & x_2, & x_3, & x_4 & \geq & 0. \end{cases} \tag{4.1}$$

This is a system of $m = 2$ equations and $n = 4$ variables. Note that the equations $x_i = 0$, $i = 1, \ldots, 4$ identify the straight lines defining the "perimeter" of the feasible region in Figure 1.3 ($x_3 = 0$ and $x_4 = 0$ identify the straight lines passing through CD and BC, respectively).

By fixing two of the 4 variables to zero we obtain the solutions coinciding with the vertices. For instance, by setting the two variables x_3 and x_4 to zero, the initial system becomes:

$$\begin{cases} 6x_1 & +4x_2 & & & = & 24 \\ 3x_1 & -2x_2 & & & = & 6 \\ & & x_3 & & = & 0 \\ & & & x_4 & = & 0 \end{cases}$$

and we obtain vertex C of Figure 1.3. Similarly, by setting x_1 and x_2 to zero we obtain vertex A, by setting x_2 and x_4 to zero we obtain vertex B, and eventually by setting x_1 and x_3 to zero we obtain vertex D. Note that $x_1 = x_4 = 0$ does not identify any vertex, as the intersection of the two corresponding straight lines lies outside the polytope P.

The considerations above can be extended using the terminology of linear algebra and matrix calculus. If there are m constraints and n variables, by setting $n - m$ variables to zero, it is possible to obtain (except for singular situations) the other m variables in a unique way. More precisely, consider the standard form in which

$$P := \{\mathbf{x} \geq 0 \ : \ A\mathbf{x} = \mathbf{b}\} \text{ with } A = [A_1, A_2, \ldots, A_n].$$

In the following, let us assume that $m \leq n$ and that A has full rank, i.e.,

$$\text{rank}(A) = m.$$

Definition 4.1.5 *A collection of m linearly independent columns of A is said to be a* basis *of A. The x_j variables associated with the basic columns are called* basic variables*; the remaining variables are called* non-basic variables.

Recalling example (4.1), we have:

$$A = \begin{bmatrix} 6 & 4 & 1 & 0 \\ 3 & -2 & 0 & 1 \end{bmatrix} , \quad \mathbf{b} = \begin{bmatrix} 24 \\ 6 \end{bmatrix} \Rightarrow B = [A_1, A_3] = \begin{bmatrix} 6 & 1 \\ 3 & 0 \end{bmatrix} \text{ basis of A.}$$

Note that basic variables can always be uniquely obtained starting from the non-basic variables. Indeed, for the sake of notation suppose that basis B is composed of the first m columns of matrix A:

$$A = [A_1, \ldots, A_n] \Rightarrow A = [B, F] \text{ with } B = [A_1, \ldots, A_m] , \quad F = [A_{m+1}, \ldots, A_n].$$

The column partition of A induces a similar partition of vector $\mathbf{x}$:

$$\mathbf{x} = \begin{bmatrix} \mathbf{x}_B \\ \mathbf{x}_F \end{bmatrix},$$

and thus the system $A\mathbf{x} = \mathbf{b}$ can be rewritten as $B\mathbf{x}_B + F\mathbf{x}_F = \mathbf{b}$.

Being B nonsingular (and thus invertible) we have:

$$B\mathbf{x}_B = \mathbf{b} - F\mathbf{x}_F \quad \Rightarrow \quad \mathbf{x}_B = B^{-1}\mathbf{b} - B^{-1}F\mathbf{x}_F,$$

and the solution

$$\mathbf{x} = \begin{bmatrix} \mathbf{x}_B \\ \mathbf{x}_F \end{bmatrix} = \begin{bmatrix} B^{-1}\mathbf{b} - B^{-1}F\mathbf{x}_F \\ \mathbf{x}_F \end{bmatrix}$$

still satisfies the given system $A\mathbf{x} = \mathbf{b}$ by construction.

Definition 4.1.6 *The solution obtained imposing* $\mathbf{x}_F = 0$ *and* $\mathbf{x}_B = B^{-1}\mathbf{b}$ *is said to be the* basic solution *associated with basis B. The basic solution (and, by extension, basis B itself) is said to be* feasible *if* $\mathbf{x}_B = B^{-1}\mathbf{b} \geq 0$.

Definition 4.1.7 *A basis B is said to be* degenerate *if* $B^{-1}\mathbf{b}$ *has one or more zero components.*

Example

Consider again example (4.1), in which

$$A = \begin{bmatrix} 6 & 4 & 1 & 0 \\ 3 & -2 & 0 & 1 \end{bmatrix} , \quad \mathbf{b} = \begin{bmatrix} 24 \\ 6 \end{bmatrix}.$$

Let us choose the basis composed of the $m = 2$ columns A_1 and A_2:

$$B = [A_1, A_2] = \begin{bmatrix} 6 & 4 \\ 3 & -2 \end{bmatrix} ; \quad F = [A_3, A_4] = \begin{bmatrix} 1 & 0 \\ 0 & 1 \end{bmatrix} ; \quad \begin{bmatrix} \mathbf{x}_B \\ \mathbf{x}_F \end{bmatrix} = \begin{bmatrix} x_1 \\ x_2 \\ x_3 \\ x_4 \end{bmatrix}.$$

$$\mathbf{x}_B = B^{-1}\mathbf{b} - B^{-1}F\mathbf{x}_F \quad \Rightarrow \quad \begin{bmatrix} x_1 \\ x_2 \end{bmatrix} = B^{-1} \begin{bmatrix} 24 \\ 6 \end{bmatrix} - B^{-1} \begin{bmatrix} 1 & 0 \\ 0 & 1 \end{bmatrix} \begin{bmatrix} x_3 \\ x_4 \end{bmatrix}$$

with $\mathbf{B}^{-1} = \begin{bmatrix} 6 & 4 \\ 3 & -2 \end{bmatrix}^{-1} = \begin{bmatrix} \frac{1}{12} & \frac{1}{6} \\ \frac{1}{8} & -\frac{1}{4} \end{bmatrix} \Rightarrow \begin{bmatrix} x_1 \\ x_2 \end{bmatrix} = \begin{bmatrix} 3 \\ \frac{3}{2} \end{bmatrix} - \begin{bmatrix} \frac{1}{12} & \frac{1}{6} \\ \frac{1}{8} & -\frac{1}{4} \end{bmatrix} \begin{bmatrix} x_3 \\ x_4 \end{bmatrix}.$

We thus have the following system, which is equivalent to the original system (4.1):

$$\begin{cases} x_1 &= 3 - \frac{1}{12}x_3 - \frac{1}{6}x_4 \\ x_2 &= \frac{3}{2} - \frac{1}{8}x_3 + \frac{1}{4}x_4 \end{cases}$$

hence the basic solution $\mathbf{x} = [3, \frac{3}{2}, 0, 0]^T$ obtained by setting x_3 and x_4 to zero is feasible.
□

In the example, the basic feasible solution obtained coincedes with vertex C of the poly-tope of Figure 1.3. This correspondence is generally valid.

Theorem 4.1.3 *A point $\mathbf{x} \in P$ is a vertex of the not empty polyhedron $P := \{\mathbf{x} \geq 0 : A\mathbf{x} = \mathbf{b}\}$ if and only if $\mathbf{x}$ is a basic feasible solution of the system $A\mathbf{x} = \mathbf{b}$.*

Proof: Let us first prove the implication "$\mathbf{x}$ is a basic feasible solution $\Rightarrow \mathbf{x}$ is a vertex".

Let

$$\mathbf{x} = [\underbrace{x_1, \ldots, x_k}_{\text{positive}}, 0, \ldots, 0]^T$$

be any basic feasible solution associated with a basis B of A, where $k \geq 0$ is the number of non-zero (i.e., strictly positive) components of $\mathbf{x}$. It follows that columns $A_1, \ldots, A_k$ must be part of B, possibly together with other columns (in case of degenerate solution). Let us assume by contradiction that $\mathbf{x}$ is not a vertex. There exist thus

$$\mathbf{y} = [y_1, \ldots, y_k, 0, \ldots, 0]^T \in P$$
$$\mathbf{z} = [z_1, \ldots, z_k, 0, \ldots, 0]^T \in P$$

with $\mathbf{y} \neq \mathbf{z}$ such that $\mathbf{x} = \lambda \mathbf{y} + (1 - \lambda)\mathbf{z}$ for any $\lambda \in (0, 1)$, which implies that $k \geq 1$.

Note that both $\mathbf{y}$ and $\mathbf{z}$ must have the last components set to zero, otherwise their convex combination cannot give $\mathbf{x}$. For the hypotheses, we then have:

$$\mathbf{y} \in P \Rightarrow A\mathbf{y} = \mathbf{b} \Rightarrow A_1 y_1 + \ldots + A_k y_k = \mathbf{b}$$
$$\mathbf{z} \in P \Rightarrow A\mathbf{z} = \mathbf{b} \Rightarrow A_1 z_1 + \ldots + A_k z_k = \mathbf{b}.$$

By subtracting the second equation from the first we obtain

$$(y_1 - z_1)A_1 + \ldots + (y_k - z_k)A_k = \alpha_1 A_1 + \ldots + \alpha_k A_k = 0,$$

where $\alpha_i := y_i - z_i$, $i = 1, \ldots, k$. Hence there exist $\alpha_1, \ldots, \alpha_k$ scalars not all zero (since $\mathbf{y} \neq \mathbf{z}$) such that $\sum_{i=1}^{k} \alpha_i A_i = 0$, thus columns $A_1, \ldots, A_k$ are linearly dependent and cannot be part of the basis B ($\Rightarrow$ contradiction).

We will now prove the implication "$\mathbf{x}$ is a vertex $\Rightarrow \mathbf{x}$ is a basic solution"; the fact that the basic solution is also feasible obviously derives from the hypothesis that $\mathbf{x} \in P$.

Writing, as before, $\mathbf{x} = [x_1, \ldots, x_k, 0, \ldots, 0]^T$ with $x_1, \ldots, x_k > 0$ and $k \geq 0$, we have that:
$$\mathbf{x} \in P \Rightarrow \mathbf{Ax} = \mathbf{b} \Rightarrow A_1 x_1 + \ldots + A_k x_k = \mathbf{b}. \tag{4.2}$$

Two cases can occur:

1. columns $A_1, \ldots, A_k$ are linearly independent (or $k = 0$): by arbitrarily selecting other $m - k$ linearly independent columns (which, as is well known, is always possible), we obtain a basis B $= [A_1, \ldots, A_k, \ldots]$ whose basic associated solution is indeed $\mathbf{x}$ (which satisfies $\mathbf{Ax} = \mathbf{b}$ and has non-basic components all equal to zero), thus concluding the proof.

2. columns $A_1, \ldots, A_k$ are linearly dependent: we will prove that this case cannot actually happen. Indeed, if the columns were linearly dependent, then there would exist $\alpha_1, \ldots, \alpha_k$ not all zero and such that
$$\alpha_1 A_1 + \ldots + \alpha_k A_k = 0, \tag{4.3}$$

The sum of (4.2) and (4.3) multiplied by $\varepsilon > 0$ would give:
$$(x_1 + \varepsilon \alpha_1) A_1 + \ldots + (x_k + \varepsilon \alpha_k) A_k = \mathbf{b}.$$

Similarly, the subtraction of (4.3) from (4.2) multiplied by ε would give:
$$(x_1 - \varepsilon \alpha_1) A_1 + \ldots + (x_k - \varepsilon \alpha_k) A_k = \mathbf{b}.$$

By defining
$$\mathbf{y} := [x_1 - \varepsilon \alpha_1, \ldots, x_k - \varepsilon \alpha_k, 0, \ldots, 0]^T$$
$$\mathbf{z} := [x_1 + \varepsilon \alpha_1, \ldots, x_k + \varepsilon \alpha_k, 0, \ldots, 0]^T,$$

we would have $\mathbf{Ay} = \mathbf{b}$ and $\mathbf{Az} = \mathbf{b}$, while choosing a sufficiently small ε we would have $\mathbf{y}, \mathbf{z} \geq 0$ and thus $\mathbf{y}, \mathbf{z} \in P$, $\mathbf{y} \neq \mathbf{z}$. But since by construction
$$\mathbf{x} = \frac{1}{2}\mathbf{y} + \frac{1}{2}\mathbf{z},$$

this would mean that vertex $\mathbf{x}$ can be expressed as the strict convex combination of two distinct points of P ($\Rightarrow$ contradiction).

$\square$

Corollary 4.1.1 *Every problem* $\min\{\mathbf{c}^T\mathbf{x} : \mathbf{Ax} = \mathbf{b} , \mathbf{x} \geq 0\}$ *defined on a polytope* $P = \{\mathbf{x} \geq 0 : \mathbf{Ax} = \mathbf{b}\} \neq \emptyset$ *has at least one optimal solution coinciding with a basic feasible solution.*

Proof: According to Theorem 4.1.2, there always exists an optimal solution coinciding with a vertex of P and thus, according to Theorem 4.1.3, with a basic feasible solution. $\square$

4.2 The Simplex Method

We have seen the equivalence between the basic feasible solutions of a linear programming problem and the vertices of P. A first idea to solve a linear programming problem consists then in enumerating all the basic solutions to the problem, identifying the feasible one that minimizes the objective function. Note that the basic solutions (feasible or non-feasible) are, at most, $\binom{n}{m} = \frac{n!}{m!(n-m)!}$. The procedure may be improved having at our disposal an efficient method to:

- verify the optimality of the current basic feasible solution;

- move from a basic feasible solution to another "adjacent" one with a better value of the objective function.

This way of proceeding can still generate, in the worst case, all basic feasible solutions of the problem, but is on average very effective.

4.2.1 Optimality Test

Let as usual $\min\{\mathbf{c}^T\mathbf{x} : \mathbf{Ax} = \mathbf{b}, \mathbf{x} \geq 0\}$ be the problem to solve, and consider a feasible basis B. The system $\mathbf{Ax} = \mathbf{b}$ may be rewritten as:

$$\mathbf{Bx_B} + \mathbf{Fx_F} = \mathbf{b} \Rightarrow \mathbf{x_B} = \mathbf{B}^{-1}\mathbf{b} - \mathbf{B}^{-1}\mathbf{Fx_F} \text{ ,where } \mathbf{B}^{-1}\mathbf{b} \geq 0. \tag{4.4}$$

By replacing (4.4) in the expression of the objective function we have:

$$\mathbf{c}^T\mathbf{x} = [\mathbf{c}_B^T, \mathbf{c}_F^T] \begin{bmatrix} \mathbf{x_B} \\ \mathbf{x_F} \end{bmatrix} = [\mathbf{c}_B^T, \mathbf{c}_F^T] \begin{bmatrix} \mathbf{B}^{-1}\mathbf{b} - \mathbf{B}^{-1}\mathbf{Fx_F} \\ \mathbf{x_F} \end{bmatrix} \Rightarrow$$

$$\mathbf{c}^T\mathbf{x} = \mathbf{c}_B^T\mathbf{B}^{-1}\mathbf{b} + (\mathbf{c}_F^T - \mathbf{c}_B^T\mathbf{B}^{-1}\mathbf{F})\mathbf{x_F} = const. + \bar{\mathbf{c}}^T\mathbf{x},$$

where $\bar{\mathbf{c}}^T$ is the objective function vector expressed as a function of the non-basic variables only.

Definition 4.2.1 *The vector*

$$\bar{\mathbf{c}}^T := \mathbf{c}^T - \mathbf{c}_B^T\mathbf{B}^{-1}\mathbf{A} = [\underbrace{\mathbf{c}_B^T - \mathbf{c}_B^T\mathbf{B}^{-1}\mathbf{B}}_{=0^T}, \mathbf{c}_F^T - \mathbf{c}_B^T\mathbf{B}^{-1}\mathbf{F}]$$

is called the reduced cost vector *with respect to basis* B.

Note that, by definition, the reduced cost of a basic variable is zero. This condition uniquely determines the reduced costs of non-basic variables.

Theorem 4.2.1 *Let* B *be a feasible basis. If* $\bar{\mathbf{c}}^T := \mathbf{c}^T - \mathbf{c}_B^T B^{-1} A \geq 0^T$, *then the basic solution associated with* B *is optimal.*

Proof: Rewriting the objective function as $\mathbf{c}^T \mathbf{x} = \mathbf{c}_B^T B^{-1} \mathbf{b} + \bar{\mathbf{c}}^T \mathbf{x}$, in the hypothesis $\bar{\mathbf{c}}^T \geq 0^T$ we have that $\mathbf{c}^T \mathbf{x} \geq \mathbf{c}_B^T B^{-1} \mathbf{b}$ for all $\mathbf{x} \geq 0$ (and thus for all $\mathbf{x} \in P$), where

$$\mathbf{c}_B^T B^{-1} \mathbf{b} = [\mathbf{c}_B^T, \mathbf{c}_F^T] \begin{bmatrix} B^{-1} \mathbf{b} \\ 0 \end{bmatrix}$$ is the value of the objective function corresponding to

the basic feasible solution associated with B. $\square$

Note however that, in presence of degeneracy, we may have negative reduced costs even at an optimal basic feasible solution. This is due to the fact that reduced costs depend on the basis B and not (directly) on the corresponding basic solution.

In the numerical example (4.1), by choosing basis $B = [A_1, A_2]$ system (4.4) becomes:

$$\begin{cases} x_1 = 3 - \frac{1}{12}x_3 - \frac{1}{6}x_4 \\ x_2 = \frac{3}{2} - \frac{1}{8}x_3 + \frac{1}{4}x_4. \end{cases} \tag{4.5}$$

By expressing the objective function $-x_1 - x_2$ as a function of the non-basic variables we have:

$$-x_1 - x_2 = -\frac{9}{2} + \frac{5}{24}x_3 - \frac{1}{12}x_4,$$

hence $\mathbf{c}_B^T B^{-1} \mathbf{b} = -\frac{9}{2}$, and the reduced costs are $\bar{c}_1 = \bar{c}_2 = 0$ (by definition), $\bar{c}_3 = \frac{5}{24}$, and $\bar{c}_4 = -\frac{1}{12}$. The basic feasible solution $[3, \frac{3}{2}, 0, 0]^T$ therefore is not optimal: increasing x_4 the objective function decreases, since $\bar{c}_4 < 0$.

4.2.2   Change of Basis

Let x_h be a non-basic variable with reduced cost $\bar{c}_h < 0$. We want variable x_h to "enter the basis", and to take the largest possible positive value—such value may however be zero in case of degeneracy. All other non-basic variables are kept equal to zero. The system

$$\mathbf{x}_B = B^{-1} \mathbf{b} - B^{-1} F \mathbf{x}_F$$

can then be rewritten as

$$\mathbf{x}_B = \bar{\mathbf{b}} - \overline{A}_h x_h,$$

where

$$\bar{\mathbf{b}} := B^{-1} \mathbf{b} = [\bar{b}_1, \dots, \bar{b}_m]^T \geq 0^T$$

while

$$\overline{A}_h := B^{-1} A_h = [\bar{a}_{1h}, \dots, \bar{a}_{mh}]^T$$

represents the column corresponding to x_h in matrix $B^{-1}F$. Indicating with

$$x_{\beta[1]}, \ldots, x_{\beta[m]}$$

the basic variables and imposing the condition $\mathbf{x}_B = [x_{\beta[1]}, \ldots, x_{\beta[m]}]^T \geq 0$, we obtain the following system of inequalities:

$$
\begin{cases}
x_{\beta[1]} = \bar{b}_1 - \bar{a}_{1h}x_h \geq 0 \\
\quad \cdots \\
x_{\beta[i]} = \bar{b}_i - \bar{a}_{ih}x_h \geq 0 \\
\quad \cdots \\
x_{\beta[m]} = \bar{b}_m - \bar{a}_{mh}x_h \geq 0
\end{cases}
\Rightarrow
\begin{cases}
\bar{a}_{1h}x_h \leq \bar{b}_1 \\
\quad \cdots \\
\bar{a}_{ih}x_h \leq \bar{b}_i \\
\quad \cdots \\
\bar{a}_{mh}x_h \leq \bar{b}_m
\end{cases}
\tag{4.6}
$$

For each condition (4.6) we have two possibilities:

- $\bar{a}_{ih} \leq 0$: the i-th condition does not imply any constraint on variable $x_h \geq 0$;

- $\bar{a}_{ih} > 0$: the condition implies $x_h \leq \bar{b}_i/\bar{a}_{ih}$.

Since all conditions in (4.6) must be satisfied at the same time, the maximum value that x_h can take may be calculated as

$$\vartheta := \min\left\{ \frac{\bar{b}_i}{\bar{a}_{ih}} \; : \; i \in \{1, \ldots, m\} \,, \, \bar{a}_{ih} > 0 \right\}. \tag{4.7}$$

When x_h takes on its limit value ϑ the objective function improves by the quantity $|\bar{c}_h| \cdot \vartheta \geq 0$.

Note that in case $\bar{a}_{ih} \leq 0$ for all $i = 1, \ldots, m$, value ϑ is indefinite, and there is no limit to the growth of the non-basic variable x_h. In this case, the objective function may reach an infinitely negative value, and thus the problem is unbounded ($\min = -\infty$). Moreover, in the degenerate case in which $\bar{b}_i = 0$ for any i with $\bar{a}_{ih} > 0$ we have $\vartheta = 0$, and consequently there is no improvement in the objective function.

Denoting by t the row with $\bar{a}_{th} > 0$ such that $\vartheta = \bar{b}_t/\bar{a}_{th}$, when imposing $x_h = \vartheta$ the basic component $x_{\beta[t]}$ becomes zero, hence it "leaves" the basis. In other words, column A_h enters the current basis B substituting column $A_{\beta[t]}$ associated with row t, according to the following scheme:

$$B = [A_{\beta[1]}, \ldots, A_{\beta[t]}, \ldots, A_{\beta[m]}] \Rightarrow \tilde{B} = [A_{\beta[1]}, \ldots, A_{\beta[t-1]}, A_h, A_{\beta[t+1]}, \ldots, A_{\beta[m]}].$$

Bases B and $\tilde{B}$ differ only by one column, and are said to be *adjacent*. It is easy to prove that if B is a basis then $\tilde{B}$ is a basis too. Indeed, by premultiplying $\tilde{B}$ by B^{-1} we obtain:

$$
B^{-1}\tilde{B} = \begin{bmatrix}
1 & 0 & \bar{a}_{1h} & 0 & 0 \\
0 & 1 & \dots & 0 & 0 \\
0 & 0 & \boxed{\bar{a}_{th}} & 0 & 0 \\
0 & 0 & \dots & 1 & 0 \\
0 & 0 & \bar{a}_{mh} & 0 & 1
\end{bmatrix} \Rightarrow det(B^{-1}\tilde{B}) = (-1)^{t+t}\bar{a}_{th} \neq 0
$$

which implies $det(\tilde{B}) \neq 0$.

In example (4.5), we have $\mathbf{c}^T\mathbf{x} = -\frac{9}{2} + \frac{5}{24}x_3 - \frac{1}{12}x_4$, and thus it is better to let variable x_4 enter the basis. However, value x_4 cannot indefinitely increase, since we have to guarantee the nonnegativity of all variables. By deciding to maintain $x_3 = 0$, system (4.5) gives:

$$
\begin{cases}
x_1 = 3 - \frac{1}{6}x_4 \geq 0 \Rightarrow & x_4 \leq 18 \\
x_2 = \frac{3}{2} + \frac{1}{4}x_4 \geq 0 \Rightarrow & \text{no restriction, since } x_4 \geq 0
\end{cases}
$$

from which we derive $x_4 \leq \vartheta := 18$. Keeping $x_3 = 0$ and increasing x_4 to its limit ($x_4 = 18$) we force variable x_1 to zero, and thus it leaves the basis.

Formalizing the procedure, we obtain the algorithm shown in Figure 4.2.

As to the choice of a variable x_h with $\bar{c}_h < 0$ entering the basis, one possibility is to select the one with maximum $|\bar{c}_h|$, or the one that maximizes the improvement $|\bar{c}_h| \cdot \vartheta$ in the objective function.

4.2.3   Update of the Inverse Matrix of the Current Basis

The simplex algorithm requires the inversion of the current basis B at each iteration. Except for the first iteration, this update can be done in a parametric way, starting from the previous basis.

We will now describe a procedure based on this idea, in which the basis inversion (and its update) occurs implicitly. The procedure manages a system of equations equivalent to the initial system $A\mathbf{x} = \mathbf{b}$, but written in a form that makes the optimality test and the basic exchange immediate.

Definition 4.2.2 *Given basis B and $z = \mathbf{c}^T\mathbf{x}$, system*

$$
\begin{cases}
\mathbf{x}_B = B^{-1}\mathbf{b} - B^{-1}F\mathbf{x}_F \\
z = \mathbf{c}_B^T B^{-1}\mathbf{b} + (\mathbf{c}_F^T - \mathbf{c}_B^T B^{-1}F)\mathbf{x}_F
\end{cases}
$$

is said to be in canonical form *with respect to* B.

Simplex Algorithm (matrix form):
  begin
    let $\beta[1], \ldots, \beta[m]$ be the indices of the columns of an initial feasible basis;
    unbounded := `false`;
    optimal := `false`;
    while (*optimal*=`false`) **and** (*unbounded*=`false`) **do**
      begin
        let $B := \left[A_{\beta[1]} | \ldots | A_{\beta[m]} \right]$ be the current feasible basis;
        compute B^{-1} and set $\mathbf{u}^T := \mathbf{c}_B^T B^{-1}$;
        compute the reduced cost $\bar{c}_h := c_h - \mathbf{u}^T A_h$ of non-basic variables x_h;
        if $\bar{c}_h \geq 0 \,\forall\, x_h$ non $-$ basic **then** *optimal* := `true`
        else
          begin
            choose a non-basic variable x_h with $\bar{c}_h < 0$;
            compute $\bar{\mathbf{b}} := B^{-1}\mathbf{b}$ and $\bar{A}_h := B^{-1}A_h$;
            if $\bar{a}_{ih} \leq 0 \,\forall i \in \{1, \ldots, m\}$ **then** *unbounded* := `true`
            else
              begin
                compute $t := \arg \min\{\bar{b}_i/\bar{a}_{ih} \,,\; i \in \{1, \ldots, m\} \,:\, \bar{a}_{ih} > 0\}$;
                update the current basis by setting $\beta[t] := h$
              end
          end
      end
  end .

Figure 4.2: Simplex Algorithm (matrix form)

The system in canonical form allows the immediate calculation of the basic variables and of the value z of the objective function, as a function of the non-basic variables. In particular, by setting to zero the non-basic variables, we obtain the components of the basic solution associated with B, and the corresponding value z of the objective function.

Let us get back to example (4.1). An initial feasible basis may be easily obtained by choosing the columns associated with slack variables x_3 and x_4.

$$1^{st} \textbf{ iteration}: \quad B = [A_3, A_4] = \begin{bmatrix} 1 & 0 \\ 0 & 1 \end{bmatrix} \quad \Rightarrow \quad \begin{cases} x_3 &= 24 - 6x_1 - 4x_2 \\ x_4 &= 6 - 3x_1 + 2x_2 \\ z &= -x_1 - x_2. \end{cases}$$

The reduced costs of the non-basic variables are negative. Let us make variable $x_h = x_1$ enter the basis. By imposing the nonnegativity conditions for the basic variables, we obtain

$$\begin{cases} x_3 = 24 - 6x_1 \geq 0 & \Rightarrow \quad x_1 \leq 4 \\ x_4 = 6 - 3x_1 \geq 0 & \Rightarrow \quad x_1 \leq 2 \end{cases} \qquad \vartheta = \min\{2, 4\} = 2$$

hence variable x_1 enters the basis and takes on value $\vartheta = 2$. In this way the basic variable x_4 is set to zero on the *pivot row* $t = 2$ (the one that determined the minimum value

ϑ), hence x_4 leaves the basis. At the same time, the objective function decreases by $|\bar{c}_h|\vartheta = 2$. To iterate the procedure, we need to update the system in canonical form, expressing the new basic variables x_3, x_1 and the objective function z as a function of the new non-basic variables x_2 and x_4.

Deriving x_1 from the pivot row, we immediately obtain, by substitution,

$$x_1 = 2 + \frac{2}{3}x_2 - \frac{1}{3}x_4 \quad \Rightarrow \quad \begin{cases} x_3 = 24 - 6\left(2 + \frac{2}{3}x_2 - \frac{1}{3}x_4\right) - 4x_2 \\ z = -\left(2 + \frac{2}{3}x_2 - \frac{1}{3}x_4\right) - x_2 \end{cases}$$

hence the system in canonical form with respect to the new basis becomes:

$$2^{nd} \textbf{ iteration}: \;\; B = [A_3, A_1] \;\; \Rightarrow \quad \begin{cases} x_3 = 12 - 8x_2 + 2x_4 \\ x_1 = 2 + \frac{2}{3}x_2 - \frac{1}{3}x_4 \\ z = -2 - \frac{5}{3}x_2 + \frac{1}{3}x_4. \end{cases}$$

Let us choose x_2 as the variable entering the basis and let us compute the maximum value ϑ for x_2:

$$\begin{cases} x_3 = 12 - 8x_2 \geq 0 \;\; \Rightarrow \;\; x_2 \leq \frac{3}{2} \Rightarrow \vartheta = \frac{3}{2} \\ x_1 = 2 + \frac{2}{3}x_2 \geq 0 \;\; \Rightarrow \;\; \text{satisfied } \forall \, x_2 \geq 0. \end{cases}$$

Hence x_2 enters the basis at level $\vartheta = \frac{3}{2}$ and x_3 leaves the basis. Deriving, as previously, x_2 from the pivot row (row 1) we obtain

$$x_2 = \frac{3}{2} - \frac{1}{8}x_3 + \frac{1}{4}x_4 \quad \Rightarrow \quad \begin{cases} x_1 = 2 + \frac{2}{3}\left(\frac{3}{2} - \frac{1}{8}x_3 + \frac{1}{4}x_4\right) - \frac{1}{3}x_4 \\ z = -2 - \frac{5}{3}\left(\frac{3}{2} - \frac{1}{8}x_3 + \frac{1}{4}x_4\right) + \frac{1}{3}x_4. \end{cases}$$

$$3^{rd} \textbf{ iteration}: \;\; B = [A_2, A_1], \;\; \begin{cases} x_2 = \frac{3}{2} - \frac{1}{8}x_3 + \frac{1}{4}x_4 \\ x_1 = 3 - \frac{1}{12}x_3 - \frac{1}{6}x_4 \\ z = -\frac{9}{2} + \frac{5}{24}x_3 - \frac{1}{12}x_4. \end{cases}$$

Variable x_4 has reduced cost $\bar{c}_4 = -\frac{1}{12} < 0$, and enters the basis. Imposing the nonnegativity of x_2 and x_1 we obtain:

$$\begin{cases} x_2 = \frac{3}{2} + \frac{1}{4}x_4 \;\; \geq \;\; 0 \;\;\; \Rightarrow \text{ satisfied } \forall \, x_4 \geq 0 \\ x_1 = 3 - \frac{1}{6}x_4 \;\; \geq \;\; 0 \;\; \Rightarrow \; x_4 \leq 18 \Rightarrow \vartheta = 18. \end{cases}$$

Hence x_4 enters the basis at level $\vartheta = 18$, while variable x_1, corresponding to the pivot row (row 2), leaves the basis. Deriving, as usual, x_4 from the pivot row and substituting it in the other rows, we obtain:

$$x_4 = 18 - 6x_1 - \frac{1}{2}x_3 \quad \Rightarrow \quad \begin{cases} x_2 = \frac{3}{2} - \frac{1}{8}x_3 + \frac{1}{4}\left(18 - 6x_1 - \frac{1}{2}x_3\right) \\ z = -\frac{9}{2} + \frac{5}{24}x_3 - \frac{1}{12}\left(18 - 6x_1 - \frac{1}{2}x_3\right) \end{cases}$$

$$4^{th} \textbf{ iteration}: \quad B = [A_2, A_4], \quad \begin{cases} x_2 = 6 - \frac{3}{2}x_1 - \frac{1}{4}x_3 \\ x_4 = 18 - 6x_1 - \frac{1}{2}x_3 \\ z = -6 + \frac{1}{2}x_1 + \frac{1}{4}x_3. \end{cases}$$

Since no reduced cost is negative, the feasible solution associated with the current basis $(x_1 = x_3 = 0, x_2 = 6, x_4 = 18, z = -6)$ is optimal, and the algorithm stops.

As an exercise, the reader shall identify in Figure 1.3 the vertices corresponding to the bases generated in the 4 iterations of the procedure.

4.2.4 Tableau Form of the Simplex Method

The aforementioned steps can be conveniently organized in a table (*tableau*).

The original system ($A\mathbf{x} = \mathbf{b}$, $z = \mathbf{c}^T\mathbf{x}$) is represented by the initial tableau:

		x_1	$\ldots$	x_n		z
row 0	0		$\mathbf{c}^T$			-1
row 1						0
$\ldots$	$\mathbf{b}$		A			$\vdots$
row m						0

Considering a basis B and partitioning A in $[B, F]$, the initial tableau can be rewritten as

	x_1	$\ldots$	x_m	x_{m+1}	$\ldots$	x_n		z
0		$\mathbf{c}_B^T$			$\mathbf{c}_F^T$			-1
								0
$\mathbf{b}$		B			F			$\vdots$
								0

We want to convert this tableau into its *canonical form* with respect to B, to reflect the system

$$\mathbf{x}_B = \underbrace{B^{-1}\mathbf{b}}_{\overline{\mathbf{b}}} - \underbrace{B^{-1}F}_{\overline{F}}\mathbf{x}_F, \quad z = \underbrace{\mathbf{c}_B^T B^{-1}\mathbf{b}}_{-\overline{c}_0} + \underbrace{(\mathbf{c}_F^T - \mathbf{c}_B^T B^{-1}F)}_{\overline{c}_F}\mathbf{x}_F.$$

To this end, we premultiply the original tableau by B^{-1}, obtaining

		x_1	$\ldots$	x_m	x_{m+1}	$\ldots$	x_n		z
$-z$	$\bar{c}_0$	0	$\ldots$	0		$\bar{c}_F^T$			-1
$x_{\beta[1]}$									0
$\vdots$	$\bar{\mathbf{b}}$		I			$\overline{\mathrm{F}}$			$\vdots$
$x_{\beta[m]}$									0

The labels associated with the tableau rows remind us of the corresponding basic variables ($-z$ for row 0, $x_{\beta[1]}$ for row 1, $x_{\beta[2]}$ for row 2, and so on).

In the tableau in canonical form, the basic columns form the identity matrix, while the reduced costs corresponding to the basic variables are all equal to zero. The basic components of the current basic solution can be read in the first column ($\bar{\mathbf{b}} := \mathrm{B}^{-1}\mathbf{b}$). The value $\bar{c}_0$ appearing in position $(0,0)$ of the tableau is the opposite of the value of the objective function calculated for the current basic solution ($-\bar{c}_0 := \mathbf{c}_B^T \mathrm{B}^{-1}\mathbf{b}$). Columns $\overline{A}_h := \mathrm{B}^{-1}A_h$ of matrix $\overline{\mathrm{F}} := \mathrm{B}^{-1}\mathrm{F}$ contain the necessary values $\bar{a}_{ih}$ to compute ϑ. Note that the column corresponding to variable z has no information (it is always equal to $[-1, 0, \ldots, 0]^T$), and may thus be omitted from the tableau.

Recalling example (4.1):

$$
\begin{cases}
\min z = & -x_1 & -x_2 & & & \\
& 6x_1 & +4x_2 & +x_3 & & = & 24 \\
& 3x_1 & -2x_2 & & +x_4 & = & 6 \\
& x_1, & x_2, & x_3, & x_4 & \geq & 0
\end{cases}
$$

we obtain the following initial tableau, already in canonical form with respect to the basis $\mathrm{B} = [A_3, A_4]$:

		x_1	x_2	x_3	x_4
$-z$	0	-1	-1	0	0
x_3	24	6	4	1	0
x_4	6	3	-2	0	1

In row 0 of the previous tableau there are some negative reduced costs. Let x_1 enter the basis, and compute

$$
\vartheta = \min\left\{ \frac{\bar{b}_i}{\bar{a}_{ih}} : \bar{a}_{ih} > 0 \right\} = \min\left\{ \frac{24}{6}, \frac{6}{3} \right\} = 2 \Rightarrow x_4 \text{ leaves.}
$$

The element with value 3, which is on the column of the variable entering the basis (x_1) and on the pivot row (row 2), is said to be the *pivot element*, and is highlighted with a circle. The following operations consist in obtaining x_1 from the pivot row, and replacing it in the remaining row. In relation to the tableau this means: divide the last row by 3, so that value 1 is highlighted in the circled position; subtract from row 1 the new row 2 multiplied by 6; and finally sum the new row 2 to row 0. This operation is called the *pivot operation* on the circled element. The tableau obtained, in canonical form with respect to the new basis, is the following:

		x_1	x_2	x_3	x_4
$-z$	0	-1	-1	0	0
x_3	24	6	4	1	0
x_4	6	③	-2	0	1
new row 2:	2	1	$-\frac{2}{3}$	0	$\frac{1}{3}$

$\Rightarrow$

		x_1	x_2	x_3	x_4
$-z$	2	0	$-\frac{5}{3}$	0	$\frac{1}{3}$
x_3	12	0	8	1	-2
x_1	2	1	$-\frac{2}{3}$	0	$\frac{1}{3}$

The corresponding basic feasible solution can be immediately read from the tableau: $z = -2$, $x_3 = 12$, $x_1 = 2$, all the other variables are equal to zero.

Let x_2 enter the basis, and choose the pivot element (with value 8) by means of the minimum ratio rule $\bar{b}_i/\bar{a}_{ih}$. With the corresponding pivot operation we divide row 1 by 8; we subtract the new row 1 from row 0 multiplied by $-\frac{5}{3}$; and finally we subtract from row 2 the new row 1 multiplied by $-\frac{2}{3}$. In this way, we obtain the new tableau:

		x_1	x_2	x_3	x_4
$-z$	2	0	$-\frac{5}{3}$	0	$\frac{1}{3}$
x_3	12	0	⑧	1	-2
x_1	2	1	$-\frac{2}{3}$	0	$\frac{1}{3}$

$\Rightarrow$

		x_1	x_2	x_3	x_4
$-z$	$\frac{9}{2}$	0	0	$\frac{5}{24}$	$-\frac{1}{12}$
x_2	$\frac{3}{2}$	0	1	$\frac{1}{8}$	$-\frac{1}{4}$
x_1	3	1	0	$\frac{1}{12}$	$\frac{1}{6}$

With similar steps we obtain:

		x_1	x_2	x_3	x_4
$-z$	$\frac{9}{2}$	0	0	$\frac{5}{24}$	$-\frac{1}{12}$
x_2	$\frac{3}{2}$	0	1	$\frac{1}{8}$	$-\frac{1}{4}$
x_1	3	1	0	$\frac{1}{12}$	⑥ $\frac{1}{6}$

$\Rightarrow$

		x_1	x_2	x_3	x_4
$-z$	6	$\frac{1}{2}$	0	$\frac{1}{4}$	0
x_2	6	$\frac{3}{2}$	1	$\frac{1}{4}$	0
x_4	18	6	0	$\frac{1}{2}$	1

procedure PIVOT(t, h);
   begin
      $save := \bar{a}[t, h]$; /* value of the pivot element */
      for $j := 0$ **to** n **do** $\bar{a}[t, j] := \bar{a}[t, j]/save$;
      for $i := 0$ **to** m **do**
         if $(i \neq t)$ **and** $(\bar{a}[i, h] \neq 0)$ **then**
            begin
               $save := \bar{a}[i, h]$;
               for $j := 0$ **to** n **do** $\bar{a}[i, j] := \bar{a}[i, j] - save * \bar{a}[t, j]$
            end
   end

Figure 4.3: Pivot operation (without variable *save*, $i > h$ would not be updated)

The tableau obtained is optimal, since there are no negative reduced costs. The corresponding optimal solution, with value $z = -6$, is given by $x_2 = 6$, $x_4 = 18$, $x_1 = x_3 = 0$. The reader may compare the procedure with that used in the previous section.

The fundamental point of the described algorithm is represented by the pivot operation on the circled element $\bar{a}_{th}$. This operation is aimed at

- bringing value 1 in position (t, h), dividing the entire row t by $\bar{a}_{th} > 0$;

- bringing value 0 in position (i, h), subtracting from row i the *new* row t multiplied by $\bar{a}_{ih}$ (for every row $i \neq t$).

The procedure che be formalized as in Figure 4.3 (Pivot Operation) and in Figure 4.4 (Simplex Algorithm).

4.2.5   The Two-Phase Method

So far we have considered a linear programming problem $\min\{\mathbf{c}^T\mathbf{x} \,:\, \mathbf{Ax} = \mathbf{b}, \mathbf{x} \geq \mathbf{0}\}$, in which $\mathbf{b} \geq \mathbf{0}$ and A contains an identity matrix of order m. This situation always occurs starting from a problem of the type:

$$\min\{\mathbf{c}^T\mathbf{x} \,:\, \mathbf{Ax} \leq \mathbf{b}, \text{ with } \mathbf{x} \geq \mathbf{0}\}, \mathbf{b} \geq \mathbf{0} \;\Rightarrow\; \min\{\mathbf{c}^T\mathbf{x} \,:\, \mathbf{Ax} + \mathbf{Is} = \mathbf{b}, \mathbf{x} \geq \mathbf{0}, \mathbf{s} \geq \mathbf{0}\}.$$

When this does not happen, we need to find an initial feasible basis (if it exists, i.e., if the problem is not infeasible). A first possibility (*Big-M method*) uses the transformation

$$\min\{\mathbf{c}^T\mathbf{x} \,:\, \mathbf{Ax} = \mathbf{b}, \mathbf{x} \geq \mathbf{0}\} \;\Rightarrow\; \min\{\mathbf{c}^T\mathbf{x} + M\sum_{i=1}^{m} y_i \,:\, \mathbf{Ax} + \mathbf{Iy} = \mathbf{b}, \mathbf{x} \geq \mathbf{0}, \mathbf{y} \geq \mathbf{0}\},$$

where $\mathbf{y}$ is a vector of "artificial" variables, M is a very large positive scalar, and $\mathbf{b} \geq \mathbf{0}$. This technique is based on the fact that, for the choice of M, at the optimum we will have $\mathbf{y} = \mathbf{0}$. However, having a very large value in the objective function may lead to significant numerical instability.

SIMPLEX Algorithm (tableau form) ;
  begin
    let $\beta[1], \ldots, \beta[m]$ be the indices of the columns of an initial feasible basis;
    create the initial tableau $\overline{A} = \{\overline{a}[i,j] \ : \ i = 0, \ldots, m \ ; \ j = 0, \ldots, n\}$, in canonical
      form with respect to the initial basis;
    unbounded := `false`;
    optimal := `false`;
    while (*optimal*=`false`) **and** (*unbounded*=`false`) **do**
      begin
        if $\overline{a}[0,j] \geq 0 \,\forall\, j = 1, \ldots, n$ **then** *optimal* := `true`
        else
          begin
            choose a non-basic variable x_h with $\overline{a}[0,h] < 0$;
            if $\overline{a}[i,h] \leq 0 \,\forall\, i = 1, \ldots, m$ **then** *unbounded* := `true`
            else
              begin
                compute $t := \arg\min \{\overline{a}[i,0]/\overline{a}[i,h] \ : \ i \in \{1, \ldots, m\}, \ \overline{a}[i,h] > 0\}$;
                PIVOT(t,h) ; /* tableau update */
                $\beta[t] := h$
              end
          end
      end
  end .

Figure 4.4: Simplex Algorithm (tableau form)

A more elegant method, called the *Two-Phase Method*, consists in applying the simplex algorithm twice: in the first phase, an initial basis is identified (if it exists), while in the second phase the problem is actually solved.

Starting from the problem in standard form $\min\{\mathbf{c}^T\mathbf{x} \ : \ A\mathbf{x} = \mathbf{b} \, , \ \mathbf{x} \geq 0\}$ with $\mathbf{b} \geq 0$ we define the *artificial problem*

$$\min\left\{ w = \sum_{i=1}^{m} y_i \ : \ A\mathbf{x} + I\mathbf{y} = \mathbf{b} \, , \ \mathbf{x} \geq 0 \, , \ \mathbf{y} \geq 0 \right\}.$$

The artificial problem has the same constraint matrix as the original problem, to which the identity matrix has been placed side by side. The *artificial variables* $y_1, \ldots, y_m$ have been added to the original variables of the problem. The artificial objective function, w, is the sum of all artificial variables.

Note that the artificial problem is already in canonical form with respect to the feasible basis associated with the m artificial variables. The objective function can be easily converted in canonical form, deriving from the system the variables y_i as a function of the non-basic variables x_j. The simplex algorithm can be hence directly applied.

Let then w^* be the optimal value of the artificial objective function, and let $(\mathbf{x}^*, \mathbf{y}^*)$ be the corresponding optimal solution. Two cases can occur:

1. $w^* > 0$. In this case, there exists no solution to the artificial problem with $y_i = 0 \; \forall \, i \in \{1, \ldots, m\}$ (such solution would have cost $w = 0$), hence the system $A\mathbf{x} = \mathbf{b}$, $\mathbf{x} \geq 0$, does not admit a solution: the original problem is hence infeasible.

2. $w^* = 0$. In this case we have $\mathbf{y}^* = 0$ (since $\mathbf{y}^* \geq 0$ and $\sum_{i=1}^{m} y_i^* = 0$), hence $\mathbf{x}^*$ is a feasible solution of the original problem. With respect to the optimal tableau for the artificial problem, two subcases may occur:

   (a) All artificial variables are non-basic variables: eliminating the columns of the tableau corresponding to the artificial variables, we have then a tableau (equivalent to the initial one) for the original problem, already in canonical form with respect to a basis. Now the artificial objective function has to be replaced with the original one, the objective function is converted in canonical form (by means of a substitution of variables), and the original problem is solved with the simplex algorithm.

   (b) There exists a basic variable y_h on a row i: in this case we have $y_h^* = 0$ (since $w^* = 0$), hence we are in the presence of degeneracy. The situation is as follows:

		x_1	$\ldots$	x_j	$\ldots$	x_n	y_1	$\ldots$	y_h	$\ldots$	y_m
$-w$	0								0		
									0		
									0		
y_h	0	$\bar{a}_{i1}$	$\ldots$	$\bar{a}_{ij}$	$\ldots$	$\bar{a}_{in}$			1		
									0		
									0		

If there exist a value $\bar{a}_{ij} \neq 0$ in correspondence to a variable x_j $(j \leq n)$, then it is sufficient to perform a pivot operation on element (i, j). Since $\bar{b}_i = 0$, this operation is admissible even if $\bar{a}_{ij} < 0$ and does not entail variations to the objective function w. In this way, y_h leaves the basis. Repeating the procedure for all the basic y_h leads us back to the non-degenerate case (a).

One last possibility remains: all values $\bar{a}_{i1}, \ldots, \bar{a}_{in}$ are equal to zero. In this case, eliminating the artificial columns (an operation that is always admissible) we obtain a tableau equivalent to the original one, but with a zero row. This implies that the matrix A does not have full row rank m, and thus the i-th row of the current tableau may be eliminated without problems (the corresponding equation of system the $A\mathbf{x} = \mathbf{b}$ is linearly dependent from the others).

Examples

In order to clarify the Two-Phase Method, consider the following LP Problems.

Feasible Problem Let us consider the following problem:

$$\begin{cases} \min z = & x_1 & & +x_3 & & \\ & x_1 & +2x_2 & & \leq & 5 \\ & & x_2 & +2x_3 & = & 6 \\ & x_1, & x_2, & x_3 & \geq & 0. \end{cases}$$

The problem does not have an obvious initial basis in the form of an identity matrix, due to the equality constraint on the second row. Adding a slack variable x_4 on the first constraint and the artificial variables y_1 and y_2 we obtain the artificial problem:

		x_1	x_2	x_3	x_4	y_1	y_2
$-w$	0	0	0	0	0	1	1
y_1	5	1	2	0	1	1	0
y_2	6	0	1	2	0	0	1

Note that the artificial variable y_1 could be omitted, since x_4 (or x_1) automatically qualifies as the basic variable in row 1. The objective function $w = y_1 + y_2$ is not in canonical form. Subtracting from row 0 the sum of the other rows, we easily obtain the complete canonical form:

		x_1	x_2	x_3	x_4	y_1	y_2
$-w$	-11	-1	-3	-2	-1	0	0
y_1	5	①	2	0	1	1	0
y_2	6	0	1	2	0	0	1

Applying the simplex algorithm, we obtain the tableaux:

		x_1	x_2	x_3	x_4	y_1	y_2
$-w$	-6	0	-1	-2	0	1	0
x_1	5	1	②	0	1	1	0
y_2	6	0	1	2	0	0	1

		x_1	x_2	x_3	x_4	y_1	y_2
$-w$	$-\frac{7}{2}$	$\frac{1}{2}$	0	-2	$\frac{1}{2}$	$\frac{3}{2}$	0
x_2	$\frac{5}{2}$	$\frac{1}{2}$	1	0	$\frac{1}{2}$	$\frac{1}{2}$	0
y_2	$\frac{7}{2}$	$-\frac{1}{2}$	0	②	$-\frac{1}{2}$	$-\frac{1}{2}$	1

		x_1	x_2	x_3	x_4	y_1	y_2
$-w$	0	0	0	0	0	1	1
x_2	$\frac{5}{2}$	$\frac{1}{2}$	1	0	$\frac{1}{2}$	$\frac{1}{2}$	0
x_3	$\frac{7}{4}$	$-\frac{1}{4}$	0	1	$-\frac{1}{4}$	$-\frac{1}{4}$	$\frac{1}{2}$

In this case $w^* = 0$ and variables y_1 and y_2 can be eliminated without problems. Moving on to phase 2, we bring back to row 0 the original objective function $z = x_1 + x_3$, obtaining:

		x_1	x_2	x_3	x_4
$-z$	0	1	0	1	0
x_2	$\frac{5}{2}$	$\frac{1}{2}$	1	0	$\frac{1}{2}$
x_3	$\frac{7}{4}$	$-\frac{1}{4}$	0	(1)	$-\frac{1}{4}$

As before, the objective function is not in canonical form (it still depends on basic variable x_3). Deriving x_3 from the last row of the tableau—or, equivalently, performing a pivot operation on the circled element—we easily obtain the canonical form:

		x_1	x_2	x_3	x_4
$-z$	$-\frac{7}{4}$	$\frac{5}{4}$	0	0	$\frac{1}{4}$
x_2	$\frac{5}{2}$	$\frac{1}{2}$	1	0	$\frac{1}{2}$
x_3	$\frac{7}{4}$	$-\frac{1}{4}$	0	1	$-\frac{1}{4}$

In this example, the tableau obtained is optimal, and further iterations of the simplex algorithm are not needed.

Infeasible Problem   Let us now consider the following problem:

$$
\begin{cases}
\min z = & x_1 & & +x_3 & & \\
& x_1 & +2x_2 & & \leq & -5 \\
& & x_2 & +2x_3 & = & 6 \\
& x_1, & x_2, & x_3 & \geq & 0
\end{cases}
$$

The problem is obviously infeasible as $x_1 + 2x_2$ cannot be negative. We nevertheless apply the Two-Phase Method. Note that it is necessary to change the sign at the first inequality, so as to have $\mathbf{b} \geq 0$ as required. The corresponding artificial problem is:

		x_1	x_2	x_3	x_4	y_1	y_2
$-w$	0	0	0	0	0	1	1
y_1	5	-1	-2	0	-1	1	0
y_2	6	0	1	2	0	0	1

Subtracting from row 0 the sum of the other rows, we obtain the canonical form:

		x_1	x_2	x_3	x_4	y_1	y_2
$-w$	-11	1	1	-2	1	0	0
y_1	5	-1	-2	0	-1	1	0
y_2	6	0	1	②	0	0	1

Performing the pivot operation on the circled element, we obtain:

		x_1	x_2	x_3	x_4	y_1	y_2
$-w$	-5	1	2	0	1	0	1
y_1	5	-1	-2	0	-1	1	0
x_3	3	0	$\frac{1}{2}$	1	0	0	$\frac{1}{2}$

In this case $w^* = 5$, and thus it is not possible to eliminate from the problem the artificial variables. It follows that the original problem is infeasible.

Artificial Basic Variables at Zero Level. Let us now consider problem:

$$\begin{cases} \min \ z = \ \ x_1 \ +x_2 \ +10x_3 \\ \qquad\qquad\quad x_2 \ +4x_3 \ = \ 2 \\ \qquad -2x_1 \ +x_2 \ -6x_3 \ = \ 2 \\ \qquad\quad x_1, \quad x_2, \quad x_3 \ \geq \ 0. \end{cases}$$

The initial tableau for phase 1 is the following:

		x_1	x_2	x_3	y_1	y_2
$-w$	-4	2	-2	2	0	0
y_1	2	0	①	4	1	0
y_2	2	-2	1	-6	0	1

With a pivot operation we obtain the optimal tableau:

		x_1	x_2	x_3	y_1	y_2
$-w$	0	2	0	10	2	0
x_2	2	0	1	4	1	0
y_2	0	(-2)	0	-10	-1	1

In this case $w^* = 0$, but y_2 is still a basic variable (at value zero). By eliminating the columns associated with variables y_1 and y_2, we would then lose a column of the identity matrix. However, by choosing as pivot any non-zero element among the elements $\bar{a}_{21}$, $\bar{a}_{22}$ and $\bar{a}_{23}$ (for example $\bar{a}_{21}$) we can easily "transfer" the column of the identity matrix into the zone of the original variables:

		x_1	x_2	x_3	y_1	y_2
$-w$	0	0	0	0	1	1
x_2	2	0	1	4	1	0
x_1	0	1	0	5	$\frac{1}{2}$	$-\frac{1}{2}$

In this way, we obtain an optimal equivalent basis, in which all basic variables are the x_j variables of the original problem (if, instead, we had had $\bar{a}_{21} = \bar{a}_{22} = \bar{a}_{23} = 0$, we would have simply eliminated row 2 from the tableau). With phase 2 we create the initial and optimal tableau:

		x_1	x_2	x_3
$-z$	-2	0	0	1
x_2	2	0	1	4
x_1	0	1	0	5

4.2.6   Convergence and Degeneracy

We have seen that the feasible bases of a problem are at most, $\binom{n}{m} = \frac{n!}{m!(n-m)!}$. Since this number is finite, albeit astronomically large in real applications, we might think that the simplex algorithm has to necessarily converge to the optimal solution in a finite number of iterations. This is certainly true if the algorithm does not pass more than once through the same basis. For example, if at all iterations the minimum ratio ϑ is strictly positive, the objective function decreases strictly monotonically, hence no basis can be visited more than once.

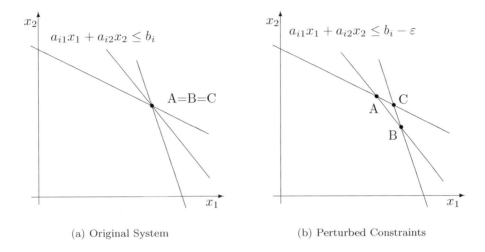

(a) Original System   (b) Perturbed Constraints

Figure 4.5: Degenerate Vertex

On the other hand, in the presence of degeneracy some values $\bar{b}_i$ are zero, and value the ϑ calculated with (4.7) can also be zero (it will not be so only if $\bar{a}_{ih} \leq 0$ for all i rows with $\bar{b}_i = 0$). The result of the pivot operation will then be a change of basis that does not change the corresponding basic feasible solution (vertex), since variable x_h enters the basis at level $\vartheta = 0$. Hence, after a certain number of degenerate iterations, there is the *concrete* risk to pass twice through a certain basis, creating thus a previously generated tableau. This circumstance may cause the simplex method to cycle indefinitely through a sequence of degenerate bases, all associated with the same vertex. In this case we talk about *cycling degeneracy*. In the example of Figure 4.5, the cycle may involve the bases associated with points A, B and C (coincident for $\varepsilon = 0$).

There are various ways to solve the problem caused by cycling degeneracy. A method used by several commercial solvers is to perturb the problem when the cycling risk occurs, i.e., when the objective function does not improve for many iterations. Perturbation consists in varying, sometimes in a pseudo-random way, the vector $\mathbf{b}$ of the right-hand sides. If the perturbation is sufficiently strong, the algorithm typically manages to get out of the degenerate point. Obviously it will then be necessary to remove the introduced perturbation.

Another possibility is to adopt anti-cycling rules for the choice of the variables that enter and leave the basis at each iteration. For example, it is possible to randomize both the choice of variable x_h (among those of negative reduced cost) and the choice of pivot row t (among those for which $\bar{b}_t/\bar{a}_{th} = \min\{\bar{b}_i/\bar{a}_{ih} : \bar{a}_{ih} > 0\}$). In this way, the probability of generating cycles tends asymptotically to zero.

One of the easiest and most elegant anti-cycling rule has been proposed by R.G. Bland: using this rule, we can demonstrate that the simplex algorithm converges in a finite number of iterations.

Bland's Rule: *Whenever it is possible to choose, always choose the entering/leaving variable x_j with the smallest index j.*

In particular, we have to:

- choose variable x_h to enter the basis defining $h := \arg\min\{j \ : \ \bar{c}_j < 0\}$;

- among all rows t with $\bar{b}_t/\bar{a}_{th} = \vartheta$ that are eligible for the pivot operation, choose the one with minimum $\beta[t]$ so as to force the smallest-index variable $x_{\beta[t]}$ to leave the basis.

As an example, consider Bland's rule applied to the following tableau:

		x_1	x_2	x_3	x_4	x_5	x_6	x_7
$-z$	-10	5	-1	0	-10	0	0	0
x_5	8	1	④	0	1	1	0	0
x_3	6	-1	③	1	0	0	0	0
x_6	1	0	-2	0	3	0	1	0
x_7	2	3	①	0	-2	0	0	1

In order to choose variable x_h to enter the basis, it is necessary to choose the negative reduced cost having minimum index (x_2). Once choice $x_h = x_2$ has been determined, the possible candidates for pivot row t are rows t with $\bar{b}_t/\bar{a}_{th} = \vartheta = 2$:

$$\text{pivot on row } t = 1 \quad \rightarrow \quad x_{\beta[1]} = x_5 \text{ leaves the basis}$$
$$\text{pivot on row } t = 2 \quad \rightarrow \quad x_{\beta[2]} = x_3 \text{ leaves the basis}$$
$$\text{pivot on row } t = 4 \quad \rightarrow \quad x_{\beta[4]} = x_7 \text{ leaves the basis.}$$

The basic variable with minimum index is x_3, hence the pivot row chosen is row $t = 2$.

Theorem 4.2.2 *Using Bland's rule, the simplex algorithm converges after, at most, $\binom{n}{m}$ iterations.*

Proof: Let us suppose by contradiction that the thesis is false and let us consider as counterexample the *smallest* LP problem for which there is no convergence. As already seen, in this case the simplex algorithm has to "go through" a cyclic sequence $B_1, B_2, \ldots, B_k = B_1$ of bases. During this sequence, pivot operations are performed on *all* rows and *all* columns of the tableau, otherwise eliminating the not involved rows/columns we would obtain a smaller counterexample. It follows that *all* variables enter and leave, in turn, the current basis. Moreover, we must have $\bar{b}_t = 0$ for all rows $t \in \{1, \ldots, m\}$, otherwise in the iteration in which the pivot operation is performed on row t, we would have $\vartheta > 0$, hence the value of the objective function would change—preventing cycling in the sequence of bases.

Consider now tableau T in which variable x_n leaves the current basis to let a given non-basic variable x_h enter the basis. Let us indicate with $x_{\beta[i]}$ the basic variable in row $i \in \{1, \ldots, m\}$, and with t the row in which x_n is in the basis (i.e., $\beta[t] = n$).

		...	$x_{\beta[i]}$	...	x_h	...	x_n	
$-z$			0		$-$		0	
...	0		0		$-$		0	...
$x_{\beta[i]}$	0		1		$-$		0	$\mu_i \geq 0$
...	0		0		$-$		0	...
$x_{\beta[t]}$	0		0		$\oplus$		1	$\mu_t < 0$
...	0		0		$-$		0	...

Tableau T :

The main features of this tableau are:

- $\bar{b}_i = 0$ $\forall i$ (as already seen)

- $\bar{c}_h < 0$ (since x_h enters the basis)

- $\bar{c}_{\beta[i]} = 0$ $\forall i$ (reduced costs basic variables)

- $\bar{a}_{th} > 0$ (pivot element)

- $\bar{a}_{ih} \leq 0$ $\forall i \neq t$ (Bland's rule).

Note that if there existed $i \neq t$ with $\bar{a}_{ih} > 0$, then Bland's rule would certainly have preferred to let variable $x_{\beta[i]}$ (instead of x_n) leave the basis.

Consider now tableau $\tilde{T}$ in which x_n re-enters the basis:

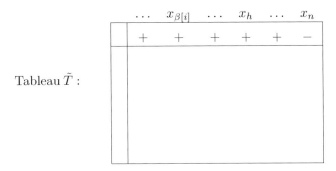

Tableau $\tilde{T}$:

Indicating with $\tilde{c}_j$ the reduced costs in row 0 of $\tilde{T}$, we must have

- $\tilde{c}_n < 0$ (since x_n enters the basis)

- $\tilde{c}_j \geq 0 \; \forall j \neq n$ (Bland's rule).

Now, $\tilde{T}$ has been obtained from T by means of a sequence of pivot operations, hence there exist appropriate multipliers $\mu_1, \ldots, \mu_m$ such that:

$$[\text{row 0 of } \tilde{T}] = [\text{row 0 of } T] + \sum_{i=1}^{m} \mu_i [\text{row } i \text{ of } T].$$

But then:

- $\underbrace{\tilde{c}_{\beta[t]}}_{=\tilde{c}_n < 0} = \underbrace{\overline{c}_{\beta[t]}}_{=0} + \mu_t \quad \Rightarrow \quad \mu_t < 0$

- $\underbrace{\tilde{c}_{\beta[i]}}_{\geq 0} = \underbrace{\overline{c}_{\beta[i]}}_{=0} + \mu_i \quad \Rightarrow \quad \mu_i \geq 0 \; \forall i \neq t$

hence there is a contradiction:

$$\underbrace{\tilde{c}_h}_{\geq 0} = \underbrace{\overline{c}_h}_{<0} + \sum_{i \neq t} \underbrace{\overline{a}_{ih}}_{\leq 0} \underbrace{\mu_i}_{\geq 0} + \underbrace{\overline{a}_{th}}_{>0} \underbrace{\mu_t}_{<0} < 0.$$

$\square$

4.3   Revised Simplex Method

The simplex algorithm in tableau form is easy to describe and implement, but is not very suitable for solving problems of medium-large size. The method indeed requires the storage and the update of a tableau of dimension $(m+1) \times (n+1)$, typically dense of non-zero coefficients—even in the case of an initially very sparse matrix A.

In practice, it is preferable to use the simplex algorithm in matrix form, outlined on page 42. The fundamental steps of this algorithm are the computation of the inverse matrix B^{-1} and of vector $\mathbf{u}^T := \mathbf{c}_B^T B^{-1}$. We will now see an easy method that allows one to iteratively update B^{-1} and $\mathbf{u}^T$ by performing a pivot operation similar to the one seen for the simplex method in tableau form.

To illustrate the process, let us imagine to extend the initial tableau adding m *auxiliary* columns according to the following scheme:

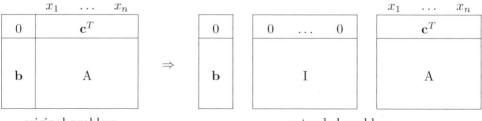

original problem   extended problem

Suppose now to apply the simplex method in tableau form to the extended problem, updating at each iteration the *entire* tableau (including the auxiliary columns) but preventing the auxiliary columns to enter the basis (i.e the pivot operations on these columns are "forbidden"). Consider a generic basis B and the corresponding extended tableau in canonical form with respect to B. As already observed, rows $1, \ldots, m$ of the tableau in canonical form with respect to B have been obtained from the original ones by premultiplication by B^{-1}, while the reduced costs in row 0 have been obtained by means of operation

$$[\text{row } 0] := [\text{original row } 0] - \sum_{i=1}^{m} u_i \, [\text{original row } i],$$

where the vector of multipliers $\mathbf{u}^T = [u_1, \ldots, u_m]$ is defined as

$$\mathbf{u}^T := \mathbf{c}_B^T B^{-1}$$

so as to have reduced costs equal to zero for the basic variables. This means that the original tableau and the one in canonical form are of the following type:

				x_1	$\cdots$	x_m	x_{m+1}	$\cdots$	x_n	
0	0	$\cdots$	0		$\mathbf{c}_B^T$			$\mathbf{c}_F^T$		
$\mathbf{b}$		I			B			F		$\Rightarrow$

original extended tableau

		x_1	$\cdots$	x_m	x_{m+1}	$\cdots$	x_n
$-\mathbf{u}^T\mathbf{b}$	$0^T - \mathbf{u}^T I = -\mathbf{u}^T$	$\mathbf{c}_B^T - \mathbf{u}^T B = 0^T$			$\mathbf{c}_F^T - \mathbf{u}^T F = \bar{\mathbf{c}}_F^T$		
$B^{-1}\mathbf{b}$	B^{-1}	I			$B^{-1}F$		

extended tableau in canonical form with respect to B

Note that $-\mathbf{u}^T$ and B^{-1} *explicitly* appear in the auxiliary columns of the tableau.

The revised simplex algorithm stores only the relevant part of the extended tableau: column 0 and the auxiliary m columns (CARRY matrix).

At each iteration, the algorithm:

1. Computes the reduced costs $\bar{c}_h := c_h - \mathbf{u}^T A_h$, and identifies the variable x_h which enters the basis;

2. Explicitly generates column $\begin{bmatrix} \bar{c}_h \\ \bar{A}_h \end{bmatrix}$, where $\bar{A}_h := B^{-1}A_h$, associated with x_h in the tableau in canonical form with respect to B, and places it temporarily side by side with the current CARRY matrix (*augmentation*);

3. Chooses the pivot element (minimum $\bar{b}_i / \bar{a}_{ih} \geq 0$), performs the pivot operation on the augmented CARRY matrix, and finally eliminates the column corresponding to variable x_h. The new CARRY matrix thus stores all relevant information about the new basis, and the procedure is iterated.

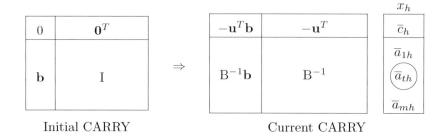

Initial CARRY   Current CARRY

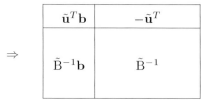

Updated (new) CARRY

As an example, consider problem (4.1), and its initial tableau:

		x_1	x_2	x_3	x_4
$-z$	0	-1	-1	0	0
x_3	24	6	4	1	0
x_4	6	3	-2	0	1

The CARRY matrices and the corresponding augmented matrices are as follows:

				x_1
$-z$	0	0	0	-1
x_3	24	1	0	6
x_4	6	0	1	③

$$\mathbf{u}^T = [0,0]$$
$$\bar{c}_1 := c_1 - \mathbf{u}^T A_1 = -1 \qquad \bar{A}_1 := \begin{bmatrix} 1 & 0 \\ 0 & 1 \end{bmatrix} \begin{bmatrix} 6 \\ 3 \end{bmatrix} = \begin{bmatrix} 6 \\ 3 \end{bmatrix}$$
$$\bar{c}_2 := c_2 - \mathbf{u}^T A_2 = -1$$

				x_2
$-z$	2	0	$\frac{1}{3}$	$-\frac{5}{3}$
x_3	12	1	-2	$\boxed{8}$
x_1	2	0	$\frac{1}{3}$	$-\frac{2}{3}$

$$\mathbf{u}^T = [0,-\tfrac{1}{3}]$$
$$\bar{c}_2 := c_2 - \mathbf{u}^T A_2 = -\frac{5}{3} \qquad \bar{A}_2 := \begin{bmatrix} 1 & -2 \\ 0 & \frac{1}{3} \end{bmatrix} \begin{bmatrix} 4 \\ -2 \end{bmatrix} = \begin{bmatrix} 8 \\ -\frac{2}{3} \end{bmatrix}$$
$$\bar{c}_4 := c_4 - \mathbf{u}^T A_4 = \frac{1}{3}$$

				x_4
$-z$	$\frac{9}{2}$	$\frac{5}{24}$	$-\frac{1}{12}$	$-\frac{1}{12}$
x_2	$\frac{3}{2}$	$\frac{1}{8}$	$-\frac{1}{4}$	$-\frac{1}{4}$
x_1	3	$\frac{1}{12}$	$\frac{1}{6}$	$\boxed{\frac{1}{6}}$

$$\mathbf{u}^T = [-\tfrac{5}{24}, \tfrac{1}{12}]$$
$$\bar{c}_3 := c_3 - \mathbf{u}^T A_3 = \frac{5}{24} \qquad \bar{A}_4 := \begin{bmatrix} \frac{1}{8} & -\frac{1}{4} \\ \frac{1}{12} & \frac{1}{6} \end{bmatrix} \begin{bmatrix} 0 \\ 1 \end{bmatrix} = \begin{bmatrix} -\frac{1}{4} \\ \frac{1}{6} \end{bmatrix}$$
$$\bar{c}_4 := c_4 - \mathbf{u}^T A_4 = -\frac{1}{12}$$

$-z$	6	$\frac{1}{4}$	0
x_2	6	$\frac{1}{4}$	0
x_4	18	$\frac{1}{2}$	1

$$\mathbf{u}^T = [-\tfrac{1}{4}, 0]$$
$$\bar{c}_1 := c_1 - \mathbf{u}^T A_1 = \frac{1}{2} \qquad \Rightarrow \text{optimal}$$
$$\bar{c}_3 := c_3 - \mathbf{u}^T A_3 = \frac{1}{4}$$

The extension of the procedure to the case in which an initial basis is not known (two-phase method) is left as an exercise for the reader.

4.4   Simplex Method for Bounded Variables

Frequently, there is the need to solve an LP problem of the type

$$\min\{\, \mathbf{c}^T\mathbf{x} \ : \ A\mathbf{x} = \mathbf{b}\,,\, \mathbf{0} \le \mathbf{x} \le \mathbf{q}\,\}$$

where $q_j \in \Re_+ \cup \{+\infty\}$ is said to be the *upper bound* of variable x_j $(j = 1, \ldots, n)$. The explicit inclusion of upper-bound constraints among the constraints of the problem would dramatically increase the number of rows of matrix A, which would go from m to $m + n$. It is therefore preferable to treat condition $\mathbf{x} \le \mathbf{q}$ only implicitly, similarly to what is done with the nonnegative constraint $\mathbf{x} \ge \mathbf{0}$.

Following such approach, given a basis B, the original matrix A is partitioned in $[B, L, U]$ where the columns of L are said to be *non-basic at lower bound* (i.e., at zero), while the columns of U are said to be *non-basic at upper bound*. Indicating with B, L and U the sets of indices of the variables that identify the submatrices B, L and U, respectively, the basic solution $(\mathbf{x}_B, \mathbf{x}_L, \mathbf{x}_U)$ associated with $[B, L, U]$ is obtained setting non-basic variables at their lower or upper bound:

$$x_j := 0 \ , \ \forall j \in L$$

$$x_j := q_j \ , \ \forall j \in U$$

and calculating basic variables as

$$\mathbf{x}_B := B^{-1}(\mathbf{b} - U\mathbf{x}_U). \tag{4.8}$$

The basis is said to be *feasible* if all basic variables x_j satisfy $0 \le x_j \le q_j$; note that $B^{-1}\mathbf{b}$ may have negative components even in a feasible basis.

Given a feasible basis, it is possible to compute the reduced cost vector $\bar{\mathbf{c}}^T = \mathbf{c}^T - \mathbf{u}^T A$ setting as usual $\mathbf{u}^T := \mathbf{c}_B^T B^{-1}$ so as to force to zero the reduced cost of all basic variables. The optimality test has to take into account the fact that non-basic variables at lower bound may only increase, while the ones at upper bound may only decrease, hence the optimality condition becomes

$$\bar{c}_j \ge 0\,, \ \forall j \in L$$

$$\bar{c}_j \le 0\,, \ \forall j \in U$$

Given a non-basic variable x_h whose reduced cost $\bar{c}_h$ does not satisfy the corresponding optimality condition, the simplex algorithm will perform a change of basis maintaining all non-basic variables other than x_h unchanged, increasing by ϑ the value of x_h (with $\vartheta \ge 0$ if x_h is non-basic at lower bound, while $\vartheta \le 0$ if x_h is non-basic at upper bound) in order to obtain the maximum improvement of the objective function.

Value ϑ can be easily determined imposing that each basic variable x_j remains within its feasibility range $[0, q_j]$ even considering the update (4.8), using formulae similar to (4.7) but which require to take into account the four possible cases depending on the sign of $\bar{a}_{ih}$ and of ϑ. Also the obvious condition $x_h \in [0, q_h]$ requiring $|\vartheta| \le q_h$ must be taken into consideration.

Once value ϑ is determined, the corresponding change of basis shall be performed which may consist in either (i) x_h entering the basis and a variable $x_{\beta[t]}$ (at its lower or upper bound) leaving the basis, or (ii) changing of the status of x_h from non-basic variable at lower bound to non-basic variable at upper bound, or vice versa. In case (i), the change of basis may request the pivot operation to be performed on a negative element $\overline{a}_{ih}$.

The implementation details of the above procedure are rather cumbersome, and are left to the interested reader.

Chapter 5

Duality in Linear Programming

Any "primal" linear programming problem in minimization form is associated with a "dual" linear programming problem in maximization form. The two problems are defined on different spaces, but have the same optimal value of the objective function (except for particular situations).

5.1 Valid Inequalities

Consider the linear programming problem in standard form $\min\{\mathbf{c}^T\mathbf{x} : \mathbf{x} \in P\}$ defined on the non-empty polyhedron $P := \{\mathbf{x} \geq 0 : A\mathbf{x} = \mathbf{b}\}$. Suppose, for the sake of simplicity, that the finite minimum exists; it is easy to see that the following *polarity relation* applies:

$$\min\{\mathbf{c}^T\mathbf{x} : x \in P\} = \max_{c_0}\{c_0 : \mathbf{c}^T\mathbf{x} \geq c_0 \; \forall \mathbf{x} \in P\}. \tag{5.1}$$

Indeed, given the values $f_1, \ldots, f_k$ of the objective function $\mathbf{c}^T\mathbf{x}$ computed in correspondence of the vertices $\mathbf{x}^1, \ldots, \mathbf{x}^k$ of P, we have:

$$\min\{f_1, f_2, \ldots, f_k\} = \max_{c_0}\{c_0 : f_i \geq c_0 , i = 1, \ldots, k\}.$$

The problem $\min\{\mathbf{c}^T\mathbf{x} : \mathbf{x} \in P\}$ consists then in identifying the maximum value c_0 for which the linear inequality $\mathbf{c}^T\mathbf{x} \geq c_0$ is verified for all vertices of P, and thus for all $\mathbf{x} \in P$. Obviously, c_0 does not exist if $\mathbf{c}^T\mathbf{x}$ is not bounded from below in P, meaning that the maximization problem in (5.1) is infeasible (min = max = $-\infty$).

Definition 5.1.1 *Given a set* $X \subseteq \Re^n$, *an inequality* $\mathbf{c}^T\mathbf{x} \geq c_0$ *satisfied by all* $\mathbf{x} \in X$ *is said to be* valid *for* X.

The maximization problem in (5.1) requires in theory to impose a constraint $\mathbf{c}^T\mathbf{x} \geq c_0$ for each $\mathbf{x} \in P$. Even considering only the vertices of P, the number of these constraints is usually astronomically large. It is therefore necessary to algebraically characterize the valid inequalities of P, answering the question: under which conditions a pair $(\mathbf{c}, c_0) \in \Re^{n+1}$ defines an inequality $\mathbf{c}^T\mathbf{x} \geq c_0$ that is valid for a given $P := \{\mathbf{x} \geq 0 : A\mathbf{x} = \mathbf{b}\}$? A partial answer (sufficient condition for validity) is based on the following simple considerations.

Let us consider a specific example:

$$A\mathbf{x} = \mathbf{b},\, \mathbf{x} \geq 0 \Rightarrow \begin{cases} 3x_1 & + & 2x_2 & - & x_3 & & & = & 5 \\ -x_1 & + & 2x_2 & + & x_3 & - & x_4 & - & 6 \\ x_1 & , & x_2 & , & x_3 & , & x_4 & \geq & 0. \end{cases}$$

We want to obtain a pair $(\mathbf{c}, c_0)$ such that the inequality $c_1x_1 + c_2x_2 + c_3x_3 + c_4x_4 \geq c_0$ is valid for P. To this end, we sum the two equations multiplied by two random scalars, u_1 and u_2, obtaining:

$$(3u_1 - u_2)x_1 + (2u_1 + 2u_2)x_2 + (-u_1 + u_2)x_3 + (-u_2)x_4 = 5u_1 + 6u_2. \tag{5.2}$$

For instance, choosing $u_1 = -1$ and $u_2 = 2$ we obtain the equation

$$-5x_1 + 2x_2 + 3x_3 - 2x_4 = 7.$$

By construction, equation (5.2) is satisfied by all $\mathbf{x} \in P$. Starting from this equation, we can obtain valid inequalities for P:

1. replacing "=" with "$\geq$";

2. reducing the right-hand side;

3. increasing the coefficient of some arbitrarily-chosen variables x_j; since $x_j \geq 0$, this operation can only increase the left-hand side.

For instance, starting from $-5x_1 + 2x_2 + 3x_3 - 2x_4 = 7$ we can obtain the valid inequality $-5x_1 + 3x_2 + 3x_3 - x_4 \geq 6$.

In general, starting from a system $Ax = b$ and choosing any $u \in \Re^m$, we have the equation

$$u^T A x = u^T b$$

from which we can obtain valid inequalities $c^T x \geq c_0$ with

$$c^T \geq u^T A \quad \text{(increase of the coefficients on the left-hand side)}$$

$$c_0 \leq u^T b \quad \text{(reduction of the right-hand side).}$$

In this way we can obtain ∞^{m+n+1} valid inequalities for P. We may wonder whether there are valid inequalities that cannot be obtained in this way. In other words: given $c^T x \geq c_0$, is the sufficient condition for validity "$\exists u \in \Re^m : c^T \geq u^T A, c_0 \leq u^T b$" also necessary? The answer to this question is of fundamental importance for duality theory.

Theorem 5.1.1 (Farkas' Lemma) *The inequality* $c^T x \geq c_0$ *is valid for the non-empty polyhedron* $P := \{x \geq 0 : Ax = b\}$ *if and only if* $u \in \Re^m$ *exists such that*

$$c^T \geq u^T A \tag{5.3}$$

$$c_0 \leq u^T b. \tag{5.4}$$

Proof: As already seen, the fact that the condition is sufficient is trivially true, given that for all $x \geq 0$ such that $Ax = b$ we have:

$$c^T x \geq u^T A x = u^T b \geq c_0.$$

We will now prove that the condition is also necessary, i.e., that "$c^T x \geq c_0$ valid for $P \neq \emptyset \Rightarrow \exists u \in \Re^m : c^T \geq u^T A, c_0 \leq u^T b$". By the hypothesis of validity, we have that

$$c_0 \leq z^* := \min\{c^T x : Ax = b, x \geq 0\}, \tag{5.5}$$

which excludes $z^* = -\infty$. Let then x^* be an optimal basic feasible solution found by the simplex algorithm applied to problem (5.5). This solution exists by the convergence property of the simplex algorithm. In addition, let B be an optimal basis associated with x^*, and let us partition as usual $A = [B, F]$, $c^T = [c_B^T, c_F^T]$, and $x^* = (x_B^*, x_F^*)$ with $x_B^* = B^{-1} b$ and $x_F^* = 0$. We will prove that vector

$$u^T := c_B^T B^{-1}$$

verifies conditions (5.3) and (5.4), hence that the thesis is valid. Recalling the reduced cost expression computed in correspondence of the optimal basis B we have:

$$\bar{c}^T := c^T - \underbrace{c_B^T B^{-1}}_{u^T} A \geq 0^T \quad \Rightarrow \quad c^T \geq u^T A$$

and thus (5.3) is valid. In addition, for (5.5) we have that

$$c_0 \leq z^* = c^T x^* = c_B^T x_B^* + c_F^T x_F^* = c_B^T B^{-1} b = u^T b,$$

and thus also (5.4) is verified. $\square$

5.2   Dual Problem

The considerations above show that, if problem $\min\{\mathbf{c}^T\mathbf{x} \ : \ A\mathbf{x} = \mathbf{b}\,,\ \mathbf{x} \geq 0\}$ has an optimal finite value, then:

$$\min_{\mathbf{x}}\{\mathbf{c}^T\mathbf{x} \ : \ A\mathbf{x} = \mathbf{b}\,,\ \mathbf{x} \geq 0\} \ = \ \max_{c_0}\{c_0 \ : \ \mathbf{c}^T\mathbf{x} \geq c_0 \text{ valid for } P\}$$

$$= \ \max_{c_0,\mathbf{u}}\{c_0 \ : \ c_0 \leq \mathbf{u}^T\mathbf{b}\,,\ \mathbf{c}^T \geq \mathbf{u}^T A\} \qquad \text{(by Farkas' lemma)}$$

$$= \ \max_{\mathbf{u}}\{\mathbf{u}^T\mathbf{b} \ : \ \mathbf{c}^T \geq \mathbf{u}^T A\} \quad \text{(since, at the optimal solution, } c_0 = \mathbf{u}^T\mathbf{b}).$$

Problems

$$\begin{cases} \min & \mathbf{c}^T\mathbf{x} \\ & A\mathbf{x} = \mathbf{b} \\ & \mathbf{x} \geq 0 \end{cases} \qquad \begin{cases} \max & \mathbf{u}^T\mathbf{b} \\ \cdot & \mathbf{c}^T \geq \mathbf{u}^T A \\ \ \end{cases}$$

are called *primal* and *dual problem*, respectively. They are characterized by the same data (matrix A and vectors $\mathbf{c}$ and $\mathbf{b}$) which, however, play different roles in the two problems. The dual problem has a constraint $c_j \geq \mathbf{u}^T A_j$ for each variable x_j of the primal problem, and a variable u_i for each constraint $\mathbf{a}_i^T\mathbf{x} = b_i$ of the primal problem. The vector $\mathbf{b}$ of right-hand sides in the primal becomes the cost vector in the dual; vice versa, the primal cost vector $\mathbf{c}$ becomes the right-hand side vector in the dual. Matrix A is transposed passing to the dual (writing $\mathbf{c}^T \geq \mathbf{u}^T A$ as $A^T\mathbf{u} \leq \mathbf{c}$). Note that the dual variables u_i are free (they are multipliers for the *equations* $\mathbf{a}_i^T\mathbf{x} = b_i$). In addition, the constraints of the dual are *inequalities* (this is due to the fact that $x_j \geq 0$ implies that the inequality $\mathbf{u}^T A\mathbf{x} \geq \mathbf{u}^T\mathbf{b}$ remains valid even increasing coefficient $\mathbf{u}^T A_j$).

Obviously, it is possible to consider the dual problem associated to a primal in canonical form:

$$\begin{cases} \min & \mathbf{c}^T\mathbf{x} \\ & A\mathbf{x} \geq \mathbf{b} \\ & \mathbf{x} \geq 0 \end{cases} \ \equiv \ \begin{cases} \min & \mathbf{c}^T\mathbf{x} + \mathbf{0}^T\mathbf{s} \\ & A\mathbf{x} - I\mathbf{s} = \mathbf{b} \\ & \mathbf{x},\mathbf{s} \geq 0 \end{cases} \ \Rightarrow \ \begin{cases} \max & \mathbf{u}^T\mathbf{b} \\ & \mathbf{c}^T \geq \mathbf{u}^T A \\ & \mathbf{0}^T \geq \mathbf{u}^T(-I) \end{cases}$$

$$\qquad\ \textit{canonical form} \qquad\qquad\ \textit{standard form} \qquad\qquad\qquad \textit{dual}$$

i.e.:

$$\begin{cases} \min & \mathbf{c}^T\mathbf{x} \\ & A\mathbf{x} \geq \mathbf{b} \\ & \mathbf{x} \geq 0 \end{cases} \ \Rightarrow \ \begin{cases} \max & \mathbf{u}^T\mathbf{b} \\ & \mathbf{c}^T \geq \mathbf{u}^T A \\ & \mathbf{u} \geq 0 \end{cases}$$

$$\qquad\qquad \textit{primal} \qquad\qquad\qquad \textit{dual}$$

Note that multipliers u_i are no longer free, since primal constraints of the type "$\geq$" had to be combined.

More generally, there are the following transformation rules, whose proof of correctness is left as an exercise to the reader:

Primal (min)	Dual (max)
$\mathbf{a}_i^T \mathbf{x} \geq b_i$	$u_i \geq 0$
$\mathbf{a}_i^T \mathbf{x} \leq b_i$	$u_i \leq 0$
$\mathbf{a}_i^T \mathbf{x} = b_i$	u_i free
$x_j \geq 0$	$\mathbf{u}^T A_j \leq c_j$
$x_j \leq 0$	$\mathbf{u}^T A_j \geq c_j$
x_j free	$\mathbf{u}^T A_j = c_j$

For example:

$$
\begin{cases}
\min & 10x_1 & +20x_2 & +30x_3 & \\
& 2x_1 & -x_2 & & \geq 1 \\
& & x_2 & +x_3 & \leq 2 \\
& x_1 & & -x_3 & = 3 \\
& x_1 & & & \geq 0 \\
& & x_2 & & \leq 0 \\
& & & x_3 & \text{free}
\end{cases}
\Rightarrow
\begin{cases}
\max & u_1 & +2u_2 & +3u_3 & \\
& u_1 & & & \geq 0 \\
& & u_2 & & \leq 0 \\
& & & u_3 & \text{free} \\
& 2u_1 & & +u_3 & \leq 10 \\
& -u_1 & +u_2 & & \geq 20 \\
& & u_2 & -u_3 & = 30
\end{cases}
$$

5.3 Fundamental Properties

An important property of the duality transformation is that it gives back the starting problem if applied twice. In other words:

Proposition 5.3.1 *The dual of the dual problem coincides with the primal problem.*

Proof: Considering without loss of generality a primal problem in canonical form and applying the known equivalence and transformation rules, we obtain:

$$
\begin{cases}
\min & \mathbf{c}^T \mathbf{x} & \\
& A\mathbf{x} & \geq & \mathbf{b} \\
& \mathbf{x} & \geq & 0
\end{cases}
\Rightarrow
\begin{cases}
\max & \mathbf{u}^T \mathbf{b} & \\
& \mathbf{c}^T \geq & \mathbf{u}^T A \\
& \mathbf{u} & \geq & 0
\end{cases}
\equiv
\begin{cases}
-\min & (-\mathbf{b}^T)\mathbf{u} & \\
& (-A^T)\mathbf{u} & \geq & -\mathbf{c} \\
& \mathbf{u} & \geq & 0
\end{cases}
$$

$$\Rightarrow \quad \begin{cases} -\max \quad \mathbf{y}^T(-\mathbf{c}) \\ \qquad\quad -\mathbf{b}^T \;\geq\; \mathbf{y}^T(-A^T) \\ \qquad\qquad\;\; \mathbf{y} \;\geq\; 0 \end{cases} \equiv \quad \begin{cases} \min \quad \mathbf{c}^T\mathbf{y} \\ \qquad\quad A\mathbf{y} \;\geq\; \mathbf{b} \\ \qquad\quad \mathbf{y} \;\geq\; 0. \end{cases}$$

$\square$

As already mentioned, we then have the following fundamental property, stated for a primal problem in canonical form:

Theorem 5.3.1 (Strong Duality) *Let* $P := \{\mathbf{x} \geq 0 \;:\; A\mathbf{x} \geq \mathbf{b}\} \neq \emptyset$ *with* $\min\{\mathbf{c}^T\mathbf{x} \;:\; \mathbf{x} \in P\}$ *finite. Then*

$$\min\{\mathbf{c}^T\mathbf{x} \;:\; A\mathbf{x} \geq \mathbf{b},\; \mathbf{x} \geq 0\} \;=\; \max\{\mathbf{u}^T\mathbf{b} \;:\; \mathbf{c}^T \geq \mathbf{u}^T A,\; \mathbf{u} \geq 0\}.$$

Proof: The proof is the result of the considerations mentioned above. $\square$

Note the the theorem does not apply when the primal problem is infeasible ($\min = +\infty$) or unbounded ($\min = -\infty$). In these cases, the following result is useful:

Theorem 5.3.2 (Weak Duality) *Let* $P := \{\mathbf{x} \geq 0 \;:\; A\mathbf{x} \geq \mathbf{b}\} \neq \emptyset$ *and* $D := \{\mathbf{u} \geq 0 \;:\; \mathbf{c}^T \geq \mathbf{u}^T A\} \neq \emptyset$. *For all pairs of points* $\overline{\mathbf{x}} \in P$ *and* $\overline{\mathbf{u}} \in D$ *we have that*

$$\overline{\mathbf{u}}^T\mathbf{b} \leq \mathbf{c}^T\overline{\mathbf{x}}.$$

Proof: Given $\overline{\mathbf{x}} \in P$ and $\overline{\mathbf{u}} \in D$, we have $A\overline{\mathbf{x}} \geq \mathbf{b}$, $\overline{\mathbf{x}} \geq 0$, $\overline{\mathbf{u}} \geq 0$ and $\mathbf{c}^T \geq \overline{\mathbf{u}}^T A$, from which

$$\overline{\mathbf{u}}^T\mathbf{b} \leq \overline{\mathbf{u}}^T A\overline{\mathbf{x}} \leq \mathbf{c}^T\overline{\mathbf{x}}.$$

$\square$

According to the previous theorem, we have that the values of the primal and dual solutions bound one another. The result is the following:

Corollary 5.3.1 *Consider the pair of primal* $\min\{\mathbf{c}^T\mathbf{x} \;:\; A\mathbf{x} \geq \mathbf{b}\,,\; \mathbf{x} \geq 0\}$ *and dual* $\max\{\mathbf{u}^T\mathbf{b} \;:\; \mathbf{c}^T \geq \mathbf{u}^T A\,,\; \mathbf{u} \geq 0\}$ *problems. There are only 4 cases:*

  1. *Both problems have a finite optimum, with* $\min\{\mathbf{c}^T\mathbf{x} \;:\; A\mathbf{x} \geq \mathbf{b}\,,\; \mathbf{x} \geq 0\} = \max\{\mathbf{u}^T\mathbf{b} \;:\; \mathbf{c}^T \geq \mathbf{u}^T A\,,\; \mathbf{u} \geq 0\}$ *;*

2. *The primal problem is unbounded and the dual is infeasible;*

3. *The dual problem is unbounded and the primal is infeasible;*

4. *Both problems are infeasible.*

Proof: Case 1 derives from the strong duality theorem. Cases 2 and 3 derive from the weak duality properties, given that the existence of a feasible solution in one of the two problems prevents the other from being unbounded. Case 4 is not excluded from any property and indeed occurs, for instance, for the following pair of problems:

$$
\begin{cases}
\min & -4x_1 & -2x_2 \\
& -x_1 & +x_2 & \geq & 2 \\
& x_1 & -x_2 & \geq & 1 \\
& x_1, & x_2 & \geq & 0
\end{cases}
\Rightarrow
\begin{cases}
\max & 2u_1 & +u_2 \\
& -u_1 & +u_2 & \leq & -4 \\
& u_1 & -u_2 & \leq & -2 \\
& u_1, & u_2 & \geq & 0.
\end{cases}
$$

$\square$

5.4 Economic Interpretation

Primal and dual problems can be interpreted economically by imagining a "competition" case between two agents in a market.

As an example, consider the following *diet problem*. A livestock farmer wants to determine the minimum-cost diet for her animals, so as to meet certain minimal nutritional requirements. On the market, there are n feedingstuffs (hay, oats, etc.), the j-th of which has a unit cost c_j. Nutritional tables consider m basic *nutrients* (proteins, vitamins, etc.). Let a_{ij} be the quantity of nutrient i present in the unit weight of feedingstuff j ($i = 1, \ldots, m, j = 1, \ldots, n$), and let b_i be the minimum nutrient quantity i that the diet has to contain ($i = 1, \ldots, m$).

The diet problem can be formulated as:

$$z^* := \min \qquad \sum_{j=1}^{n} c_j x_j$$

$$\sum_{j=1}^{n} a_{ij} x_j \geq b_i \ , \ i = 1, \ldots, m$$

$$x_j \geq 0 \ , \ j = 1, \ldots, n$$

where variable x_j gives the quantity of feedingstuff j in the diet.

The corresponding dual problem

$$p^* := \max \qquad \sum_{i=1}^{m} u_i b_i$$

$$\sum_{i=1}^{m} a_{ij} u_i \leq c_j \ , \ j = 1, \ldots, n$$

$$u_i \geq 0 \ , \ i = 1, \ldots, m$$

can be interpreted as follows.

A pharmaceutical company wants to enter the market to sell its products to livestock farmers. The company can *directly* synthesize each of the m nutrients, and has to decide the unit selling price u_i for each nutrient so as to maximize its profit. If the livestock farmer decides to use the synthetic nutrients instead of the natural feedingstuffs, she will obviously buy b_i units of each nutrient i. This explains the objective function of the dual problem, which expresses precisely the overall profit $\sum_{i=1}^{m} u_i b_i$ of the pharmaceutical company. Of course, if the prices u_i are too high, the farmer will prefer to buy the natural feedingstuffs, and therefore it is necessary for the prices u_i to be competitive with respect to that of the feedingstuffs. This condition is expressed by the constraints of the dual that set a *sufficient* condition to be competitive: the cost $\sum_{i=1}^{m} a_{ij} u_i$ to synthetically reproduce each feedingstuff j must not exceed the cost c_j of the feedingstuff itself. In this way, for any feasible choice $\bar{\mathbf{x}}$ of the farmer and for any price policy $\bar{\mathbf{u}}$ we will certainly have $\sum_{i=1}^{m} \bar{u}_i b_i \leq \sum_{j=1}^{n} c_j \bar{x}_j$ (weak duality), i.e., the farmer will surely prefer the synthetic nutrients.

The pharmaceutical company is therefore sure to be able to earn at least p^*, the optimal value of the dual problem. Is it possible to earn even more? The answer is no, given that by strong duality $z^* = p^*$. Indeed, let $\mathbf{x}^*$ be an optimal solution of the primal problem. If the pharmaceutical company decided to increase prices in order to make a profit $\tilde{p} := \sum_{i=1}^{m} u_i b_i > p^*$, the farmer would prefer to meet her needs by purchasing x_j^* units of each natural feedingstuff at a cost $\sum_{j=1}^{n} c_j x_j^* = z^* = p^* < \tilde{p}$.

The market tends towards an equilibrium point between the two competitors, in which the buyer has two options of the same cost.

5.5   Optimality Conditions

Consider again a primal-dual pair in canonical form $\min\{\mathbf{c}^T \mathbf{x} \ : \ A\mathbf{x} \geq \mathbf{b}, \ \mathbf{x} \geq 0\}$ and $\max\{\mathbf{u}^T \mathbf{b} \ : \ \mathbf{c}^T \geq \mathbf{u}^T A, \ \mathbf{u} \geq 0\}$. From the considerations above, we have that two vectors $\bar{\mathbf{x}} \in \Re^n$ and $\bar{\mathbf{u}} \in \Re^m$ are optimal for the corresponding problems if and only if:

(1) $A\bar{\mathbf{x}} \geq \mathbf{b}, \ \bar{\mathbf{x}} \geq 0$ (primal feasibility)

(2) $\mathbf{c}^T \geq \bar{\mathbf{u}}^T A, \ \bar{\mathbf{u}} \geq 0$ (dual feasibility)

(3) $\mathbf{c}^T \bar{\mathbf{x}} = \bar{\mathbf{u}}^T \mathbf{b}$ (equality of the values of the objective function: min = max).

Condition (3) imposes an equality between two scalar values. Together with conditions (1)-(2) it implies, however, a nonlinear system of $n + m$ equations in $n + m$ variables $\overline{x}_1, \ldots, \overline{x}_n$ and $\overline{u}_1, \ldots, \overline{u}_m$, obtained as follows.

By (1) and (2) we have
$$\overline{\mathbf{u}}^T \mathbf{b} \leq \overline{\mathbf{u}}^T A \overline{\mathbf{x}} \leq \mathbf{c}^T \overline{\mathbf{x}},$$
hence (3) implies $\overline{\mathbf{u}}^T \mathbf{b} = \overline{\mathbf{u}}^T A \overline{\mathbf{x}}$ and $\overline{\mathbf{u}}^T A \overline{\mathbf{x}} = \mathbf{c}^T \overline{\mathbf{x}}$, i.e.:

$$(3.\alpha)\ \overline{\mathbf{u}}^T (A \overline{\mathbf{x}} - \mathbf{b}) = 0$$

$$(3.\beta)\ (\mathbf{c}^T - \overline{\mathbf{u}}^T A) \overline{\mathbf{x}} = 0.$$

Writing $(3.\alpha)$ in full and recalling that $\overline{u}_i \geq 0$ and $\mathbf{a}_i^T \overline{\mathbf{x}} - b_i \geq 0\ \forall i$, we obtain:

$$\sum_{i=1}^{m} \overline{u}_i (\mathbf{a}_i^T \overline{\mathbf{x}} - b_i) = 0 \ \Rightarrow\ \begin{cases} \overline{u}_1(\mathbf{a}_1^T \overline{\mathbf{x}} - b_1) &= 0 \\ \quad \ldots & \\ \overline{u}_m(\mathbf{a}_m^T \overline{\mathbf{x}} - b_m) &= 0. \end{cases} \tag{5.6}$$

Proceeding in the same way, from $(3.\beta)$ we obtain:

$$\sum_{j=1}^{n} (c_j - \overline{\mathbf{u}}^T A_j) \overline{x}_j = 0 \ \Rightarrow\ \begin{cases} (c_1 - \overline{\mathbf{u}}^T A_1) \overline{x}_1 &= 0 \\ \quad \ldots & \\ (c_n - \overline{\mathbf{u}}^T A_n) \overline{x}_n &= 0. \end{cases} \tag{5.7}$$

Conditions (5.6) and (5.7) are called the *complementary conditions* and are interpreted as follows. Let $\overline{\mathbf{x}}$ and $\overline{\mathbf{u}}$ be the optimal solutions of the primal and dual problems, respectively:

- the dual variable $\overline{u}_i$ corresponding to a "loose" constraint $(\mathbf{a}_i^T \overline{\mathbf{x}} > b_i)$ must be equal to zero;

- the reduced cost $\overline{c}_j := c_j - \overline{\mathbf{u}}^T A_j$ corresponding to a variable $\overline{x}_j > 0$ must be equal to zero.

Hence we have the important

Theorem 5.5.1 (Optimality Conditions) *Two vectors $\overline{\mathbf{x}} \in \Re^n$ and $\overline{\mathbf{u}} \in \Re^m$ are optimal for problems $\min\{\mathbf{c}^T \mathbf{x} : A\mathbf{x} \geq \mathbf{b}, \mathbf{x} \geq 0\}$ and $\max\{\mathbf{u}^T \mathbf{b} : \mathbf{c}^T \geq \mathbf{u}^T A, \mathbf{u} \geq 0\}$, respectively, if and only if the following optimality conditions hold:*

$$(1)\quad A\overline{\mathbf{x}} \geq \mathbf{b}, \ \overline{\mathbf{x}} \geq 0 \qquad \text{(primal feasibility)}$$

$$(2)\quad \mathbf{c}^T \geq \overline{\mathbf{u}}^T A, \ \overline{\mathbf{u}} \geq 0 \qquad \text{(dual feasibility)}$$

$$\left.\begin{aligned} (3.\alpha)\quad & \overline{\mathbf{u}}^T (A\overline{\mathbf{x}} - \mathbf{b}) = 0 \\ (3.\beta)\quad & (\mathbf{c}^T - \overline{\mathbf{u}}^T A)\overline{\mathbf{x}} = 0 \end{aligned}\right\} \quad \text{(complementary slackness conditions)}.$$

5.6   Analysis of the Simplex Algorithm

It is interesting to interpret the simplex algorithm in the light of the optimality conditions just obtained. Since the simplex considers the problem in standard form $\min\{\mathbf{c}^T\mathbf{x}$: $A\mathbf{x} = \mathbf{b}$, $\mathbf{x} \geq 0\}$, the optimality conditions become:

$$
\begin{aligned}
&(1') \qquad A\overline{\mathbf{x}} \;=\; \mathbf{b}\,,\; \overline{\mathbf{x}} \geq 0 && \text{(primal feasibility)} \\
&(2') \qquad\quad \mathbf{c}^T \geq \overline{\mathbf{u}}^T A && \text{(dual feasibility)} \\
&(3') \quad (\mathbf{c}^T - \overline{\mathbf{u}}^T A)\overline{\mathbf{x}} \;=\; 0 && \text{(complementary conditions)},
\end{aligned}
$$

while condition $(3.\alpha)$ has been eliminated as implied by $(1')$. At each iteration, the simplex algorithm identifies a feasible basis B of matrix $A = [B, F]$ and computes the following vectors:

- $\overline{\mathbf{x}} \;=\; \left(B^{-1}\mathbf{b}, 0\right) \geq 0$

- $\overline{\mathbf{u}}^T := \mathbf{c}_B^T B^{-1}$

- $\overline{\mathbf{c}}^T := \mathbf{c}^T - \overline{\mathbf{u}}^T A,$

stopping when $\overline{\mathbf{c}} \geq 0$.

By construction, at each iteration the current basic solution $\overline{\mathbf{x}} \geq 0$ satisfies the primal feasibility condition $(1')$, since $A\overline{\mathbf{x}} = B\overline{\mathbf{x}}_B + F\overline{\mathbf{x}}_F = BB^{-1}\mathbf{b} + 0 = \mathbf{b}$. The same applies to condition $(3')$, given that:

$$
(\mathbf{c}^T - \overline{\mathbf{u}}^T A)\overline{\mathbf{x}} \;=\; \underbrace{(\mathbf{c}_B^T - \overline{\mathbf{u}}^T B)}_{\mathbf{c}_B^T - \mathbf{c}_B^T B^{-1}B = 0^T}\, \overline{\mathbf{x}}_B + (\mathbf{c}_F^T - \overline{\mathbf{u}}^T F)\underbrace{\overline{\mathbf{x}}_F}_{=0} \;=\; 0.
$$

Instead, dual feasibility $(2')$ is obtained only at the last iteration, when all reduced costs are greater than or equal to zero $(\overline{\mathbf{c}}^T := \mathbf{c}^T - \mathbf{u}^T A \geq 0^T)$.

Summarizing, at each iteration the simplex algorithm defines $\overline{\mathbf{x}}$ such that $(1')$ is satisfied, computes $\overline{\mathbf{u}}^T$ such that $(3')$ is satisfied, and performs a test to verify whether $(2')$ is satisfied. At the last iteration $(2')$ is verified, hence $\overline{\mathbf{u}}$ becomes feasible for the dual problem; this guarantees the optimality of $\overline{\mathbf{x}}$ and $\overline{\mathbf{u}}$ for the respective problems.

super-optimal solution

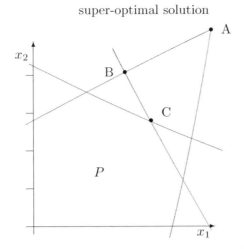

Figure 5.1: Sequence of "super-optimal" solutions

5.7 Dual Simplex Algorithm

An alternative ("dual") resolution algorithm can be obtained defining $\overline{\mathbf{x}}$ and $\overline{\mathbf{u}}$ so as to verify optimality conditions $(2')$ and $(3')$ at each iteration, while condition $(1')$ is only met at the last iteration. Geometrically, the method starts from an infeasible "super-optimal" (if it were feasible, it would be optimal) basic solution $\overline{\mathbf{x}}$, and proceeds iteratively approaching the feasible region; see Figure 5.1.

Let us now describe in detail the procedure, known as the *dual simplex algorithm*, in its tableau form. Obviously, also the "matrix" and "revised" version of the algorithm exist, the formalization of which is left as an exercise to the reader.

To avoid any confusion, the simplex algorithm described in the previous chapter is called *primal simplex*. However, it is important to see that *both* algorithm are designed to solve the *primal* problem $\min\{\mathbf{c}^T\mathbf{x} : \mathbf{A}\mathbf{x} = \mathbf{b}, \mathbf{x} \geq 0\}$.

The dual simplex algorithm needs an initial tableau in canonical form of the type:

		x_1	$\ldots$	x_t	$\ldots$	x_m	x_{m+1}	$\ldots$	x_h	$\ldots$	x_n
$-z$	$\overline{c}_0$	0	$\ldots$	0	$\ldots$	0	$\overline{c}_{m+1}$	$\ldots$	$\overline{c}_h$	$\ldots$	$\overline{c}_n$
x_1	$\overline{b}_1$	1	$\ldots$	0	$\ldots$	0	$\overline{a}_{1,m+1}$	$\ldots$		$\ldots$	$\overline{a}_{1n}$
$\ldots$	$\ldots$	0		0		0	$\ldots$	$\ldots$		$\ldots$	
x_t	$\overline{b}_t$	0		1		0	$\overline{a}_{t,m+1}$	$\ldots$	$\overline{a}_{th}$	$\ldots$	$\overline{a}_{tn}$
$\ldots$	$\ldots$	0		0		0	$\ldots$	$\ldots$		$\ldots$	
x_m	$\overline{b}_m$	0	$\ldots$	0	$\ldots$	1	$\overline{a}_{m,m+1}$	$\ldots$		$\ldots$	$\overline{a}_{mn}$

Namely, it is necessary that the reduced costs $\bar{c}_1, \ldots, \bar{c}_n$ in row 0 are all greater than or equal to zero; see the dual feasibility condition (2').

If values $\bar{b}_1, \ldots, \bar{b}_m$ are all greater than or equal to zero, the tableau is optimal. Suppose instead that a value $\bar{b}_t < 0$ exists. There are two possibilities:

- $\bar{a}_{tj} \geq 0 \;\forall j = 1, \ldots, n$: in this case, for all $\mathbf{x} \geq 0$ we will have $\sum_{j=1}^{n} \bar{a}_{tj} x_j \geq 0$, hence the equation associated with the t-th row of the tableau cannot be satisfied: the *primal* problem is infeasible.

- a term $\bar{a}_{th} < 0$ exists: in this case, by performing a pivot operation in position (t, h) we make the new term $\bar{b}_t$ positive.

Obviously, the pivot operation will also update the values in row 0, obtaining the new values $\tilde{c}_j$ computed as follows:

$$\tilde{c}_0 := \bar{c}_0 - \frac{\bar{c}_h}{\bar{a}_{th}} \bar{b}_t$$

$$\tilde{c}_j := \bar{c}_j - \frac{\bar{c}_h}{\bar{a}_{th}} \bar{a}_{tj} \;, \quad j = 1, \ldots, n.$$

If we want to maintain dual feasibility, no reduced cost $\tilde{c}_j$ ($j = 1, \ldots, n$) shall become negative, and thus we must impose

$$\bar{c}_j \geq \frac{\bar{c}_h}{\bar{a}_{th}} \bar{a}_{tj} = \frac{\bar{c}_h}{-|\bar{a}_{th}|} \bar{a}_{tj} \;\text{(given that } \bar{a}_{th} < 0) \;, \quad j = 1, \ldots, n.$$

If $\bar{a}_{tj} \geq 0$ the inequality is surely verified, given that by hypothesis $\bar{c}_j \geq 0$ and $\bar{c}_h \geq 0$. If instead $\bar{a}_{tj} < 0$, then

$$\bar{c}_j \geq \frac{\bar{c}_h}{-|\bar{a}_{th}|} (-|\bar{a}_{tj}|) \quad \Rightarrow \quad \frac{\bar{c}_j}{|\bar{a}_{tj}|} \geq \frac{\bar{c}_h}{|\bar{a}_{th}|}$$

applies.

It is therefore not possible to arbitrarily choose the pivot element $\bar{a}_{th} < 0$ in row t, but this term has to minimize the ratio $\bar{c}_j / |\bar{a}_{tj}|$ for all values $\bar{a}_{tj} < 0$, i.e.:

$$h := \arg \min \left\{ \frac{\bar{c}_j}{|\bar{a}_{tj}|} \;:\; j \in \{1, \ldots, n\}, \bar{a}_{tj} < 0 \right\}.$$

As a result of the pivot operation, the value of the objective function computed in correspondence of the current basic solution changes from $z = -\bar{c}_0$ to $\tilde{z} = -\tilde{c}_o$, hence it increases (i.e., *gets worse*) by the quantity

$$\tilde{z} - z = \bar{c}_0 - \tilde{c}_0 = \bar{c}_h \frac{\bar{b}_t}{\bar{a}_{th}} = \bar{c}_h \left| \frac{\bar{b}_t}{\bar{a}_{th}} \right| \geq 0.$$

In this way, except in the case of *dual degeneracy* in which $\bar{c}_h = 0$, the current "super-optimal" basic solution gets worse at each iteration, converging after no more than $\binom{n}{m}$ iterations to the optimal solution.

In case of dual degeneracy, the convergence can be ensured by applying anti-cycling rules as, for instance, the following

Bland's Rule (dual): *Whenever it is possible to choose, always choose the entering/leaving variable x_j with the smallest index j.*

In particular, we will have to:

1. choose variable $x_{\beta[t]}$ leaving the basis with minimum index $\beta[t]$, i.e., the pivot row t with $\bar{b}_t < 0$ and minimum $\beta[t]$.

2. among all variables x_h with $\bar{a}_{th} < 0$ and $\bar{c}_h/|\bar{a}_{th}| = \min\{\bar{c}_j/|\bar{a}_{tj}| \; : \; \bar{a}_{tj} < 0\}$ that are eligible for entering the basis, choose the one with minimum index h.

For instance, consider the following tableau:

		x_1	x_2	x_3	x_4	x_5	x_6	x_7
$-z$	-5	0	6	0	10	0	5	0
x_5	-10	0	8	0	3	1	7	0
x_3	-5	0	-1	1	(-2)	0	-1	0
x_1	3	1	-1	0	-2	0	0	0
x_7	-1	0	3	0	6	0	-2	1

We have to choose a pivot row t with $\bar{b}_t < 0$:

$$\text{pivot on row } t = 1 \quad \rightarrow \quad x_{\beta[1]} = x_5 \text{ leaves the basis}$$
$$\text{pivot on row } t = 2 \quad \rightarrow \quad x_{\beta[2]} = x_3 \text{ leaves the basis}$$
$$\text{pivot on row } t = 4 \quad \rightarrow \quad x_{\beta[4]} = x_7 \text{ leaves the basis.}$$

Since we have to make variable $x_{\beta[t]}$ with minimum index $\beta[t]$ leave the basis, we choose row $t = 2$. The elements $\bar{a}_{th} < 0$ eligible for the pivot are those that minimize the ratio $\bar{c}_h/|\bar{a}_{th}|$, i.e., the elements in position $(2,4)$ and $(2,6)$ – performing the pivot on element $(2,2)$ would not be correct: we would loose the dual feasibility given that ratio $\bar{c}_2/|\bar{a}_{22}|$ is not the minimum. We then choose as pivot element $\bar{a}_{24} = -2$, making variable x_4 with minimum index enter the basis.

5.7.1   Example

Consider the following linear programming problem:

$$
\begin{cases}
\min & 3x_1 & +4x_2 & +5x_3 & & \\
& 2x_1 & +2x_2 & +x_3 & \geq & 6 \\
& x_1 & +2x_2 & +3x_3 & \geq & 5 \\
& x_1, & x_2, & x_3 & \geq & 0.
\end{cases}
$$

The initial tableau in canonical form does not give a feasible basis: by performing the primal simplex algorithm, it would be necessary to proceed with phase 1 to identify an initial feasible basis. By changing sign to rows 1 and 2, we can instead easily obtain an initial tableau for the dual simplex algorithm:

		x_1	x_2	x_3	x_4	x_5
$-z$	0	3	4	5	0	0
x_4	-6	(-2)	-2	-1	1	0
x_5	-5	-1	-2	-3	0	1

By choosing to make x_4 leave the basis (pivot on row $t = 1$), the pivot element $\bar{a}_{th} < 0$ is determined computing the minimum ratio $\bar{c}_h/|\bar{a}_{th}|$ among:

$$
\frac{\bar{c}_1}{|\bar{a}_{11}|} = \frac{3}{2} \; ; \; \frac{\bar{c}_2}{|\bar{a}_{12}|} = 2 \; ; \; \frac{\bar{c}_3}{|\bar{a}_{13}|} = 5 \; .
$$

Hence the pivot element is $\bar{a}_{11}$ and the variable entering the basis is $x_h = x_1$. By performing the corresponding pivot operation, we obtain the new tableau:

		x_1	x_2	x_3	x_4	x_5
$-z$	-9	0	1	$\frac{7}{2}$	$\frac{3}{2}$	0
x_1	3	1	1	$\frac{1}{2}$	$-\frac{1}{2}$	0
x_5	-2	0	(-1)	$-\frac{5}{2}$	$-\frac{1}{2}$	1

By performing the pivot operation on the element in position (2,2), we obtain the feasible (and hence optimal) tableau for the primal:

		x_1	x_2	x_3	x_4	x_5
$-z$	-11	0	0	1	1	1
x_1	1	1	0	-2	-1	1
x_2	2	0	1	$\frac{5}{2}$	$\frac{1}{2}$	-1

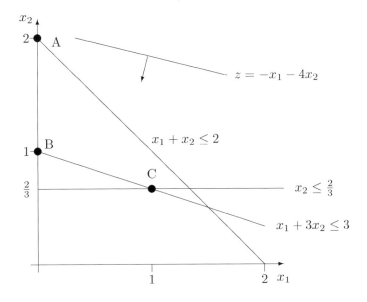

Figure 5.2: Graphical representation of the introduction of new constraints

The objective function has changed from $z = 0$ (initial tableau) to $z = 9$ (second tableau), arriving at last to the optimal value $z = 11$.

5.7.2 Introduction of additional constraints

The dual simplex algorithm is particularly useful when, after having solved a linear programming problem, we want to add new constraints. This can occur when the constraints of the primal problem are too many to be dealt with all at once, or when such constraints are not known "a priori" but can be derived starting from an optimal solution of the less constrained problem – these concept will be clarified in the chapter on integer linear programming.

Consider an optimal tableau, to which we want to add a new constraint $\alpha^T \mathbf{x} \leq \alpha_0$. By adding a slack variable, we obtain equation $\alpha^T \mathbf{x} + x_{n+1} = \alpha_0$ that can be added to the tableau. By eliminating from such equation the basic variables (by substitution) we can easily convert the new tableau in canonical form, and thus apply the dual simplex algorithm.

Let us now see an easy example. Consider the following problem, graphically represented in Figure 5.2:

$$\begin{cases} \min & -x_1 & -4x_2 \\ & x_1 & +x_2 & \leq & 2 \\ & x_1 & +3x_2 & \leq & 3 \\ & & x_2 & \leq & \frac{2}{3} \\ & x_1, & x_2 & \geq & 0. \end{cases}$$

Leaving aside for the moment the last two explicit constraints, we obtain the following tableau:

		x_1	x_2	x_3
$-z$	0	-1	-4	0
x_3	2	1	①	1

By performing the primal simplex algorithm, we reach the optimum after a single pivot operation:

		x_1	x_2	x_3
$-z$	8	3	0	4
x_2	2	1	1	1

The optimal basic solution $\mathbf{x}^* = [0,2,0]^T$ corresponds to point A in Figure 5.2. This point, however, violates both constraints left aside (i.e., it is "super-optimal" for the original problem). Adding slack variables x_4 and x_5, it is easy to turn these two constraints into equations. Let us now insert the two equations in the current tableau.

		x_1	x_2	x_3	x_4	x_5
$-z$	8	3	0	4	0	0
x_2	2	1	①	1	0	0
x_4	3	1	3	0	1	0
x_5	$\frac{2}{3}$	0	1	0	0	1

The resulting tableau can easily be converted to canonical form by means of the indicated pivot operation (eliminating basic variable x_2), obtaining

		x_1	x_2	x_3	x_4	x_5
$-z$	8	3	0	4	0	0
x_2	2	1	1	1	0	0
x_4	-3	-2	0	-3	1	0
x_5	$-\frac{4}{3}$	-1	0	-1	0	1

The basic solution associated with this tableau is still $x_1 = x_3 = 0$, $x_2 = 2$, and the negative values of the new basic variables $x_4 = -3$ and $x_5 = -\frac{4}{3}$ express the fact that this solution violates the constraints that have just been added. By construction, the new tableau can be processed with the dual simplex algorithm, given that all reduced costs are nonnegative.

By performing the pivot operation on element -3, we finally obtain the tableau associated with the new "super-optimal" point $\mathbf{x} = \left[0, 1, 1, 0, -\frac{1}{3}\right]^T$, namely point B in Figure 5.2:

		x_1	x_2	x_3	x_4	x_5
$-z$	4	$\frac{1}{3}$	0	0	$\frac{4}{3}$	0
x_2	1	$\frac{1}{3}$	1	0	$\frac{1}{3}$	0
x_3	1	$\frac{2}{3}$	0	1	$-\frac{1}{3}$	0
x_5	$-\frac{1}{3}$	$\left(-\frac{1}{3}\right)$	0	0	$-\frac{1}{3}$	1

By performing the pivot operation on value $-\frac{1}{3}$, we obtain the optimal tableau associated with the optimal vertex $\mathbf{x} = \left[1, \frac{2}{3}, \frac{1}{3}, 0, 0\right]^T$ — point C in Figure 5.2:

		x_1	x_2	x_3	x_4	x_5
$-z$	$\frac{11}{3}$	0	0	0	1	1
x_2	$\frac{2}{3}$	0	1	0	0	1
x_3	$\frac{1}{3}$	0	0	1	-1	2
x_1	1	1	0	0	1	-3

5.8 Sensitivity Analysis

Once an optimal solution for the starting problem is found, it is often interesting to evaluate the "stability" of this solution with respect to changes in the problem data. In real applications, indeed, the mathematical model is often an approximation of the real situation. The model is usually deemed more reliable if its solutions are less sensitive to changes in the data (often obtained with measurements affected by error).

Consider a problem in standard form of the type $\min\{\mathbf{c}^T\mathbf{x} : A\mathbf{x} = \mathbf{b}, \mathbf{x} \geq 0\}$ and let B be an optimal basis—e.g., the one identified by means of the simplex algorithm. Under the usual notation, the corresponding optimal basic solution is $\overline{\mathbf{x}} = (\overline{\mathbf{x}}_B, \overline{\mathbf{x}}_F)$, with $\overline{\mathbf{x}}_F = 0$ and $\overline{\mathbf{x}}_B = B^{-1}\mathbf{b}$. By theorem 5.5.1, $\overline{\mathbf{x}}$ is optimal if and only if there exists $\overline{\mathbf{u}} \in \Re^m$ such that the optimality conditions $(1') - (3')$ are satisfied, for instance $\overline{\mathbf{u}} = \mathbf{c}_B^T B^{-1}$. Note that, in case of degeneracy, $\overline{\mathbf{u}}$ is not necessarily unique, but it depends on the basis B associated with $\overline{\mathbf{x}}$. The optimality conditions written for basis B become:

$(c.1)$ $B^{-1}\mathbf{b} \geq 0$ (primal feasibility for $\overline{\mathbf{x}}$)

$(c.2)$ $\overline{\mathbf{c}}^T := \mathbf{c}^T - \underbrace{\mathbf{c}_B^T B^{-1}}_{\overline{\mathbf{u}}^T} A \geq 0^T$ (dual feasibility for $\overline{\mathbf{u}}$)

while, as is well known, the complementary condition $(3')$ derives from the choice $\overline{\mathbf{u}} = \mathbf{c}_B^T B^{-1}$.

Sensitivity analysis is the study of the perturbations to the initial data whereby conditions $(c.1)$ and $(c.2)$ remain verified. This study makes it possible to define the necessary and sufficient conditions such that basis B remains feasible and optimal when changing data: i.e., the conditions refer to *basis* B and not to the corresponding *basic solution* $\overline{\mathbf{x}}$. For the sake of simplicity, we will only consider the following possibilities:

- Changes in the right-hand sides: $\mathbf{b} \to \mathbf{b} + \Delta\mathbf{b}$

- Changes in the costs of basic variables: $\mathbf{c}_B^T \to \mathbf{c}_B^T + \Delta\mathbf{c}_B^T$

- Changes in the costs of non-basic variables: $\mathbf{c}_F^T \to \mathbf{c}_F^T + \Delta\mathbf{c}_F^T$.

Changes in the right-hand sides

Assuming a change $\Delta\mathbf{b}$ of vector $\mathbf{b}$ of the right-hand sides, the optimality conditions $(c.1)$ - $(c.2)$ for basis B become

$(c.1)$ $B^{-1}(\mathbf{b} + \Delta\mathbf{b}) \geq 0$

$(c.2)$ $\overline{\mathbf{c}}^T := \mathbf{c}^T - \mathbf{c}_B^T B^{-1} A \geq 0^T$ (unchanged).

Thus, basis B remains feasible and optimal if and only if:

$$B^{-1}\mathbf{b} \geq -B^{-1}\Delta\mathbf{b}.$$

This system of m inequalities in the m variables Δb_i defines a polyhedron in $\Re^m$ containing vectors $\Delta\mathbf{b}$ for which the optimal basis does not change. Note that when $\mathbf{b}$ changes, the coordinates of the corresponding basic solution $\overline{\mathbf{x}}$ and the optimal value of the objective function change. Indeed, the optimal value changes from $\mathbf{c}_B^T B^{-1}\mathbf{b}$ to $\mathbf{c}_B^T B^{-1}(\mathbf{b} + \Delta\mathbf{b})$, with a change of

$$\Delta z := (\mathbf{c}_B^T B^{-1})\Delta\mathbf{b} = \overline{\mathbf{u}}^T \Delta\mathbf{b} = \sum_{i=1}^{m} \overline{u}_i \Delta b_i.$$

The dual variables $\overline{u}_i$ measure thus the "sensitivity" of the optimal value of the objective function with respect to changes Δb_i of the right-hand sides.

Changes in the costs of non-basic variables

Consider now a change $\Delta\mathbf{c}_F^T$ of vector $\mathbf{c}_F^T$, and let $\bar{\mathbf{c}}$ and $\tilde{\mathbf{c}}$ be the reduced cost vectors before and after change $\Delta\mathbf{c}_F$, respectively. Conditions $(c.1)$ and $(c.2)$ become:

$(c.1)$ $\mathrm{B}^{-1}\mathbf{b} \geq 0$ (unchanged)

$(c.2)$ $\tilde{\mathbf{c}}^T := [\tilde{\mathbf{c}}_B^T, \tilde{\mathbf{c}}_F^T] = [0^T, (\mathbf{c}_F^T + \Delta\mathbf{c}_F^T) - \mathbf{c}_B^T\mathrm{B}^{-1}\mathrm{F}] \geq 0^T$

hence basis B remains optimal if and only if

$$\tilde{\mathbf{c}}_F^T = \underbrace{\mathbf{c}_F^T - \mathbf{c}_B^T\mathrm{B}^{-1}\mathrm{F}}_{\bar{\mathbf{c}}_F^T} + \Delta\mathbf{c}_F^T = \bar{\mathbf{c}}_F^T + \Delta\mathbf{c}_F^T \geq 0 \;\Leftrightarrow\; \Delta\mathbf{c}_F \geq -\bar{\mathbf{c}}_F.$$

In this way, we obtain the $n - m$ inequalities, independent from each other,

$$\Delta c_j \geq -\bar{c}_j \;\forall x_j \text{ non-basic.}$$

It follows that the reduced cost $\bar{c}_j \geq 0$ can be interpreted as the maximum *decrease* in cost c_j under which basis B remains optimal: greater decreases would produce a reduced cost $\tilde{c}_j < 0$, hence *basis* B would be no longer optimal (in case of degeneracy, however, the *basic solution* $\bar{\mathbf{x}}$ could still remain optimal).

Changes in the costs of basic variables

Consider now a change $\Delta\mathbf{c}_B^T$ of vector $\mathbf{c}_B^T$. Indicating as before with $\bar{\mathbf{c}}$ and $\tilde{\mathbf{c}}$ the reduced cost vectors before and after the change, respectively, we obtain the following conditions:

$(c.1)$ $\mathrm{B}^{-1}\mathbf{b} \geq 0$ (unchanged)

$(c.2)$ $\tilde{\mathbf{c}}^T := [\tilde{\mathbf{c}}_B^T, \tilde{\mathbf{c}}_F^T] = [0^T, \mathbf{c}_F^T - (\mathbf{c}_B^T + \Delta\mathbf{c}_B^T)\mathrm{B}^{-1}\mathrm{F}] \geq 0^T,$

from which we obtain:

$$\tilde{\mathbf{c}}_F^T := \underbrace{\mathbf{c}_F^T - \mathbf{c}_B^T\mathrm{B}^{-1}\mathrm{F}}_{\bar{\mathbf{c}}_F^T} - \Delta\mathbf{c}_B^T\mathrm{B}^{-1}\mathrm{F} \geq 0^T,$$

i.e.

$$\Delta\mathbf{c}_B^T\mathrm{B}^{-1}\mathrm{F} \leq \bar{\mathbf{c}}_F^T.$$

This system defines a polyhedron in $\Re^m$, whose points correspond to the vectors $\Delta\mathbf{c}_B$ for which the optimal basis does not change.

5.8.1   Example

Starting from the initial tableau:

		x_1	x_2	x_3	x_4	x_5	x_6
$-z$	0	-3	-1	-3	0	0	0
x_4	2	2	1	1	1	0	0
x_5	5	1	2	3	0	1	0
x_6	6	2	2	1	0	0	1

we arrive after some pivot operations to the optimal tableau

		x_1	x_2	x_3	x_4	x_5	x_6
$-z$	$\frac{27}{5}$	0	$\frac{7}{5}$	0	$\frac{6}{5}$	$\frac{3}{5}$	0
x_1	$\frac{1}{5}$	1	$\frac{1}{5}$	0	$\frac{3}{5}$	$-\frac{1}{5}$	0
x_3	$\frac{8}{5}$	0	$\frac{3}{5}$	1	$-\frac{1}{5}$	$\frac{2}{5}$	0
x_6	4	0	1	0	-1	0	1

Since the initial tableau contains the identity matrix, the inverse matrix of the optimal basis B can be directly read in the last three columns of the final tableau. The necessary pieces of information for the sensitivity analysis are:

$$
\mathbf{x_B} = \begin{bmatrix} x_1 \\ x_3 \\ x_6 \end{bmatrix} ; \quad
\mathbf{B} = \begin{bmatrix} 2 & 1 & 0 \\ 1 & 3 & 0 \\ 2 & 1 & 1 \end{bmatrix} ; \quad
\mathbf{B}^{-1} = \begin{bmatrix} \frac{3}{5} & -\frac{1}{5} & 0 \\ -\frac{1}{5} & \frac{2}{5} & 0 \\ -1 & 0 & 1 \end{bmatrix} ; \quad
\mathbf{B}^{-1}\mathbf{b} = \begin{bmatrix} \frac{1}{5} \\ \frac{8}{5} \\ 4 \end{bmatrix}
$$

$$
\mathbf{c_B} = \begin{bmatrix} -3 \\ -3 \\ 0 \end{bmatrix} ; \quad
\mathbf{x_F} = \begin{bmatrix} x_2 \\ x_4 \\ x_5 \end{bmatrix} ; \quad
\mathbf{c_F} = \begin{bmatrix} -1 \\ 0 \\ 0 \end{bmatrix} ; \quad
\mathbf{\bar{u}}^T = \mathbf{c_B}^T \mathbf{B}^{-1} = \begin{bmatrix} -\frac{6}{5}, & -\frac{3}{5}, & 0 \end{bmatrix} .
$$

Note that, in this case, values $-\bar{u}_i$ $(i = 1, 2, 3)$ coincide with the last three elements of row zero of the final tableau.

As regards a change $\Delta\mathbf{b}$ of the vectors of the right-hand sides, basis B remains optimal if and only if $B^{-1}\Delta\mathbf{b} \geq -B^{-1}\mathbf{b}$, i.e.:

$$
\begin{bmatrix} \frac{3}{5} & -\frac{1}{5} & 0 \\ -\frac{1}{5} & \frac{2}{5} & 0 \\ -1 & 0 & 1 \end{bmatrix}
\begin{bmatrix} \Delta b_1 \\ \Delta b_2 \\ \Delta b_3 \end{bmatrix}
\geq -
\begin{bmatrix} \frac{1}{5} \\ \frac{8}{5} \\ 4 \end{bmatrix}
\Rightarrow
\begin{cases}
\frac{3}{5}\Delta b_1 & -\frac{1}{5}\Delta b_2 & & \geq & -\frac{1}{5} \\
-\frac{1}{5}\Delta b_1 & +\frac{2}{5}\Delta b_2 & & \geq & -\frac{8}{5} \\
-\Delta b_1 & & +\Delta b_3 & \geq & -4.
\end{cases}
$$

Note that $\Delta\mathbf{b} = 0$ is a feasible solution to the system. Assuming, for the sake of simplicity, a change to only one component b_i at a time, we obtain:

$$
\begin{aligned}
-\tfrac{1}{3} \leq \Delta b_1 \leq 4 \quad &\text{when} \quad \Delta b_2 = \Delta b_3 = 0 \\
-4 \leq \Delta b_2 \leq 1 \quad &\text{when} \quad \Delta b_1 = \Delta b_3 = 0 \\
-4 \leq \Delta b_3 \quad\quad &\text{when} \quad \Delta b_1 = \Delta b_2 = 0.
\end{aligned}
$$

As regards change $\Delta\mathbf{c}_F$ of the reduced cost vector of non-basic variables, basis B remains optimal if and only if $\Delta\mathbf{c}_F \geq -\bar{\mathbf{c}}_F$, i.e.:

$$
\begin{bmatrix} \Delta c_2 \\ \Delta c_4 \\ \Delta c_5 \end{bmatrix}
\geq -
\begin{bmatrix} \frac{7}{5} \\ \frac{6}{5} \\ \frac{3}{5} \end{bmatrix}
\Rightarrow
\begin{cases}
\Delta c_2 & \geq & -\frac{7}{5} \\
\Delta c_4 & \geq & -\frac{6}{5} \\
\Delta c_5 & \geq & -\frac{3}{5}.
\end{cases}
$$

Let us consider a change $\Delta\mathbf{c}_B$ of the costs of the basic variables. Basis B remains optimal if and only if $\Delta\mathbf{c}_B^T B^{-1} F \leq \bar{\mathbf{c}}_F^T$. In our case, we have:

$$
\Delta\mathbf{c}_B^T = [\Delta c_1, \Delta c_3, \Delta c_6]
$$

and, directly from the optimal tableau:

$$
\bar{\mathbf{c}}_F =
\begin{bmatrix} \frac{7}{5} \\ \frac{6}{5} \\ \frac{3}{5} \end{bmatrix}
\; ; \; \overline{F} := B^{-1}F =
\begin{bmatrix} \frac{1}{5} & \frac{3}{5} & -\frac{1}{5} \\ \frac{3}{5} & -\frac{1}{5} & \frac{2}{5} \\ 1 & -1 & 0 \end{bmatrix}.
$$

We immediately obtain

$$
[\Delta c_1, \Delta c_3, \Delta c_6]
\begin{bmatrix} \frac{1}{5} & \frac{3}{5} & -\frac{1}{5} \\ \frac{3}{5} & -\frac{1}{5} & \frac{2}{5} \\ 1 & -1 & 0 \end{bmatrix}
\leq \begin{bmatrix} \frac{7}{5}, & \frac{6}{5}, & \frac{3}{5} \end{bmatrix},
$$

i.e.

$$
\begin{cases}
\frac{1}{5}\Delta c_1 & +\frac{3}{5}\Delta c_3 & +\Delta c_6 & \leq & \frac{7}{5} \\
\frac{3}{5}\Delta c_1 & -\frac{1}{5}\Delta c_3 & -\Delta c_6 & \leq & \frac{6}{5} \\
-\frac{1}{5}\Delta c_1 & +\frac{2}{5}\Delta c_3 & & \leq & \frac{3}{5}
\end{cases}
$$

In particular we will have:

$$-3 \leq \Delta c_1 \leq 2 \quad \text{when} \quad \Delta c_3 = \Delta c_6 = 0$$
$$-6 \leq \Delta c_3 \leq \tfrac{3}{2} \quad \text{when} \quad \Delta c_1 = \Delta c_6 = 0$$
$$-\tfrac{6}{5} \leq \Delta c_6 \leq \tfrac{7}{5} \quad \text{when} \quad \Delta c_1 = \Delta c_3 = 0.$$

Chapter 6

Integer Linear Programming

An *Integer Linear Programming* (ILP) problem, in canonical form, is defined as:

$$\begin{cases} z_{PLI} := \min & \mathbf{c}^T \mathbf{x} \\ & \mathbf{A}\mathbf{x} \geq \mathbf{b} \\ & \mathbf{x} \geq 0 \quad \text{integer,} \end{cases}$$

where z_{PLI} is the optimal value of the objective function.

Note that the integrality constraint on variables x_j $(j = 1, \ldots, n)$ is nonlinear; indeed, it can be mathematically expressed as

$$\sin(\pi x_j) = 0 , \quad j = 1, \ldots, n.$$

In some cases, the integrality constraint regards only some variables of the model: in such case, we talk about *Mixed-Integer Linear Programming*.

The integrality constraint on the variables defines a lattice of points in $\Re^n$, a subset of which (those satisfying $\mathbf{x} \geq 0$, $\mathbf{A}\mathbf{x} \geq \mathbf{b}$) define the feasible region of the ILP problem; see Figure 6.1 for an illustration in $\Re^2$.

In the following, we will assume that $\mathbf{A}$ and $\mathbf{b}$ are integer. We will also assume that the polyhedron

$$P := \{\mathbf{x} \geq 0 : \mathbf{A}\mathbf{x} \geq \mathbf{b}\}$$

is bounded and non-empty, and we will indicate with

$$X := P \cap Z^n$$

the (discrete and finite) set of feasible solutions of the ILP problem.

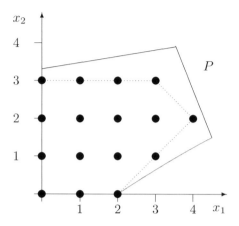

Figure 6.1: Sets P, X and $\text{conv}(X)$

6.1   Equivalent Formulations

A procedure to try to solve an ILP problem is the following:

- We eliminate the integrality constraint and solve the linear programming problem thus obtained, called the *continuous relaxation* of the original problem:

$$
\left\{
\begin{aligned}
z_{PL} := \min \quad & \mathbf{c}^T \mathbf{x} \\
& \mathbf{Ax} \geq \mathbf{b} \\
& \mathbf{x} \geq 0,
\end{aligned}
\right.
$$

  where with z_{PL} we indicate the optimal value of the continuous relaxation. Note that $X \subset P$ implies $z_{PL} := \min\{\mathbf{c}^T\mathbf{x} \ : \ \mathbf{x} \in P\} \leq \min\{\mathbf{c}^T\mathbf{x} \ : \ \mathbf{x} \in X\} =: z_{PLI}$, hence z_{PL} is a lower bound for z_{PLI}.

- If the solution $\mathbf{x}^*$ of the continuous relaxation is integer, then it is also optimal for the integer problem, as we have: $\mathbf{c}^T\mathbf{x}^* = z_{PL} \leq z_{PLI}$ but also $\mathbf{c}^T\mathbf{x}^* \geq z_{PLI}$ given that $\mathbf{x}^* \in X$, i.e. $\mathbf{c}^T\mathbf{x}^* = z_{PLI}$.

Of course, the procedure described cannot guarantee to find an optimal $\mathbf{x}^*$ solution for the ILP problem. Consider the example of Figure 6.1, in which polytope P is represented with the continuous line. According to the "slope" of the objective function, it is possible that the solution $\mathbf{x}^*$ of the continuous relaxation coincides with an integer vertex (for instance, $\mathbf{x}^* = [0,0]^T$ or $\mathbf{x}^* = [2,0]^T$) or with a non-integer vertex. As can be seen, this situation depends on the *formulation* $\mathbf{Ax} \geq \mathbf{b}$ chosen to represent the problem: using an alternative formulation $\tilde{\mathbf{A}}\mathbf{x} \geq \tilde{\mathbf{b}}$ it is possible to obtain a relaxation $\tilde{P} := \{\mathbf{x} \geq 0 \ : \ \tilde{\mathbf{A}}\mathbf{x} \geq \tilde{\mathbf{b}}\}$ (represented with the dotted line in Figure 6.1) in which *all* vertices are integer.

These simple considerations show a fundamental aspect of ILP: since the formulation of the problem is not unique, it is possible to describe the same problem by means of more or less "tight" (i.e., strong) constraints. The resulting formulations are equivalent while the integrality constraint is maintained, but they have continuous relaxations which give lower bounds z_{PL} very different from each other. Theoretically, there also exists an *ideal formulation* $\tilde{A}x \geq \tilde{b}$ (the "tightest" possible) whose relaxation $\tilde{P}$ has only integer vertices, thus guarantying $z_{PL} = z_{PLI}$ for all linear objective functions.

Definition 6.1.1 *Given a set $S \subseteq \Re^n$, a convex hull of S is the smallest convex set* conv(S) *that contains S.*

It is possible to prove that conv(X) is a polytope $\tilde{P}$ whose vertices are all integer points. There exist thus $\tilde{A}$ and $\tilde{b}$ such that $\tilde{P} := \{x \geq 0 \; : \; \tilde{A}x \geq \tilde{b}\} = $ conv(X), hence guarantying

$$\min\{c^T x \; : \; x \in X\} \; = \; \min\{c^T x \; : \; \tilde{A}x \geq \tilde{b}, \, x \geq 0\}$$

for all cost vectors c. Each ILP problem is then equivalent to an LP problem, hence it can be solved, for instance, using the simplex algorithm. Unfortunately $\tilde{A}$ and $\tilde{b}$ are often very difficult to determine and they define a system $\tilde{A}x \geq \tilde{b}$ containing a huge number of constraints, hence this solution strategy is impractical.

6.2 Total Unimodularity

In some cases, the original formulation of an ILP problem coincides with the ideal formulation. Let us consider, for the sake of simplicity, an ILP problem in standard form:

$$\begin{cases} z_{PLI} := \min & c^T x \\ & Ax = b \\ & x \geq 0 \quad \text{integer} \end{cases}$$

where, as already stated, A and b are assumed to be integer. We want to determine the conditions under which a generic basic feasible solution x^* of system $Ax = b$ has fractional components. Indicating with B the basis associated to x^*, we have:

$$x^* = \begin{bmatrix} B^{-1}b \\ 0 \end{bmatrix},$$

hence x^* has fractional components if and only if $B^{-1}b$ is noninteger. Now, matrix B^{-1} may be computed as:

$$B^{-1} = \frac{1}{det(B)} \begin{bmatrix} \alpha_{11} & \cdots & \alpha_{1m} \\ \vdots & & \vdots \\ \alpha_{m1} & \cdots & \alpha_{mm} \end{bmatrix}^T,$$

where $\alpha_{ij} := (-1)^{i+j} det(M_{ij})$ and M_{ij} is the submatrix obtained from B eliminating row i and column j. Since B is integer, the values α_{ij} are all integer. It follows that B^{-1} is integer if $det(B) = \pm 1$. In this case, $B^{-1}\mathbf{b}$ is integer as well, and the basic solution $\mathbf{x}^*$ cannot have fractional components.

Note that condition $det(B) = \pm 1$ is only sufficient, but not necessary, for the integrality of $B^{-1}\mathbf{b}$ (consider the case $det(B) = 2$ with $\mathbf{b}$ vector of even elements). However it is possible to prove that, given a matrix A containing a basis B with $|det(B)| \neq 1$, there always exists al LP problem of the kind $\min\{\mathbf{c}^T\mathbf{x} \ : \ A\mathbf{x} = \mathbf{b} \, , \, \mathbf{x} \geq 0\}$, with $\mathbf{c}$ and $\mathbf{b}$ properly defined, whose optimal solution coincides with a fractional vertex.

Definition 6.2.1 *An integer matrix* A *of size* $m \times n$ *with* $m \leq n$ *is said to be* unimodular *if for any of its* $m \times m$ *submatrices* B*,* $det(B) \in \{-1, 0, 1\}$.

Theorem 6.2.1 *Let* A *be unimodular and* $\mathbf{b}$ *integer. Then polyhedron* $P := \{\mathbf{x} \geq 0 \ : \ A\mathbf{x} = \mathbf{b}\}$ *has only integer vertices.*

Dim: Let $\mathbf{x}^*$ be any vertex of P. As already proven, there exists a basis B of A such that $\mathbf{x}^* = (B^{-1}\mathbf{b}, 0)$. Since B is a nonsingular $m \times m$ submatrix of A, we have $|det(B)| = 1$, which implies the integrality of B^{-1} and of $B^{-1}\mathbf{b}$ for all integer $\mathbf{b}$. $\square$

Consider now an ILP problem in canonical form of the kind $\min\{\mathbf{c}^T\mathbf{x} \ : \ A\mathbf{x} \geq \mathbf{b}, \mathbf{x} > 0\}$. This problem may be reformulated in standard form by adding slack variables. We obtain thus problem

$$\min\{\mathbf{c}^T\mathbf{x} \ : \ A\mathbf{x} - I\mathbf{s} = \mathbf{b} \, , \, \mathbf{x} \geq 0 \, , \, \mathbf{s} \geq 0\}$$

in which the constraint matrix becomes $A' := [A, -I]$. Every submatrix $m \times m$ B of A' is obtained selecting k (say) columns from A, and $m - k$ columns from $-I$ ($0 \leq k \leq m$). Barring permutations of rows and columns—operations that in any case do not alter $|det(B)|$—we have that B has the following form

$$B \ = \ \left[\begin{array}{c|c} -I' & F \\ \hline 0 & Q \end{array} \right]$$

$$\underbrace{}_{\text{columns from } -I} \quad \underbrace{}_{\text{columns from } A}$$

where I' is the identity matrix of order $m - k$. We therefore have that $det(B) = \pm det(Q)$, hence A' is unimodular if and only if $det(Q) \in \{-1, 0, 1\}$ for *all* square submatrices Q of A, of any order. These considerations justify the following definition:

Definition 6.2.2 *An integer matrix* A *of size* $m \times n$ *is said to be* totally unimodular *(TUM) if* $det(Q) \in \{-1, 0, 1\}$ *for any of its square submatrices* Q *of any order.*

Theorem 6.2.2 *Let* A *be totally unimodular and* $\mathbf{b}$ *integer. Then polyhedron* $P := \{\mathbf{x} \geq 0 \ : \ A\mathbf{x} \geq \mathbf{b}\}$ *has only integer vertices.*

Figure 6.2: Conditions of Theorem 6.2.3

Dim: Let $\mathbf{x}^*$ be any vertex of polyhedron P.

First of all, we prove that $(\mathbf{x}^*, \mathbf{s}^* := A\mathbf{x}^* - \mathbf{b})$ is a vertex of polyhedron

$$P' := \{(\mathbf{x}, \mathbf{s}) \geq 0 \ : \ A\mathbf{x} - \mathbf{s} = \mathbf{b}\}.$$

If that was not the case, there would exist two distinct points $(\mathbf{x}^1, \mathbf{s}^1)$ and $(\mathbf{x}^2, \mathbf{s}^2)$ of P' such that $(\mathbf{x}^*, \mathbf{s}^*) = \lambda(\mathbf{x}^1, \mathbf{s}^1) + (1 - \lambda)(\mathbf{x}^2, \mathbf{s}^2)$ for some λ with $0 < \lambda < 1$. Note that $\mathbf{x}^1$ and $\mathbf{x}^2$ belong to P, given that $\mathbf{s}^1 = A\mathbf{x}^1 - \mathbf{b} \geq 0$ and $\mathbf{s}^2 = A\mathbf{x}^2 - \mathbf{b} \geq 0$. In addition $(\mathbf{x}^1, \mathbf{s}^1) \neq (\mathbf{x}^2, \mathbf{s}^2)$ implies $\mathbf{x}^1 \neq \mathbf{x}^2$, hence $\mathbf{x}^* = \lambda\mathbf{x}^1 + (1 - \lambda)\mathbf{x}^2$ cannot be a vertex of P.

Since A is TUM, we have that $A' := [A, -I]$ is unimodular. By Theorem 6.2.1, $(\mathbf{x}^*, \mathbf{s}^*)$ is integer, hence $\mathbf{x}^*$ is integer as well. $\qquad \square$

Consider now the problem of determining whether a given integer matrix A is TUM. A simple algorithm to solve this problem consists in enumerating all submatrices Q of A, but it would require an excessive computing time even for matrices A of small size. There exist, however, more sophisticated algorithms that efficiently solve the same problem.

Some simple conditions are now defined to check if a given matrix A is TUM.

An obvious necessary condition is that we have $a_{ij} \in \{-1, 0, 1\}$ for any element (1×1 submatrix) of A. This condition is, however, not sufficient: consider, for instance, the case in which A contains the submatrix

$$Q = \begin{bmatrix} 1 & 1 \\ -1 & 1 \end{bmatrix} \text{ with } det(Q) = 2.$$

Instead, an important sufficient (but not necessary) condition is the following.

Theorem 6.2.3 *Let* A *be a matrix with* $a_{ij} \in \{-1, 0, 1\} \ \forall i, j$. A *is totally unimodular if the following conditions hold (see Figure 6.2):*

(1)   every column of A *has no more that two non-zero elements;*

(2)   there exists a partition (I_1, I_2) *of the rows of* A *such that each column with two non-zero elements has these two elements belonging to rows on different sets if and only if the two elements have the same sign.*

Proof: We have to prove that $det(Q) \in \{-1, 0-1\}$ for any submatrix Q of A of order k $(k = 1, \ldots, m)$. The proof is by induction on k.

If $k=1$ then $Q = [a_{ij}]$ and hence $det(Q) = a_{ij} \in \{-1, 0, 1\}$, as requested.

Let us suppose now that $det(Q') \in \{-1, 0, 1\}$ for any submatrix Q' of order k', where $k' \geq 1$ is a fixed value. Let us consider any submatrix Q of order $k := k'+1$. By condition (1) only three cases can occur:

- Q has one column of zeros: in this case $det(Q) = 0$.

- Q has a column with only one element different from zero: in this case, barring permutations of rows and/or columns, Q is of the type

$$Q = \begin{bmatrix} \pm 1 & * & \ldots & * \\ \hline 0 & & & \\ \vdots & & Q' & \\ 0 & & & \end{bmatrix} \quad \Rightarrow \quad det(Q) = \pm 1 \cdot det(Q')$$

where $det(Q') \in \{-1, 0, 1\}$ given that Q' has order $k - 1 = k'$.

- Each column of Q has exactly two non-zero elements: in this case, let $I(Q)$ be the set of rows of Q. By hypothesis (2), Q has the following form

$$Q = \begin{array}{l} \left. \begin{array}{cccc} 1 & & 1 & \\ & -1 & -1 & \\ \hline 1 & & & -1 \\ & -1 & & 1 \end{array} \right\} \begin{array}{l} I_1 \cap I(Q) \\ \\ I_2 \cap I(Q) \end{array} \end{array}$$

But then we have

$$\sum_{i \in I_1 \cap I(Q)} [\text{row } i \text{ of } Q] - \sum_{i \in I_2 \cap I(Q)} [\text{row } i \text{ of } Q] = [\text{null row}]$$

given that in every column the two non-zero elements cancel out. It follows that the rows of Q are linearly dependent, hence $det(Q) = 0$.

In each of the three cases above we therefore have $det(Q) \in \{-1, 0, 1\}$, hence this property applies to all submatrices Q of order $k = k' + 1$. Applying the same reasoning, the result can be inductively extended to matrices Q of any order. $\square$

The proof of the following properties is straightforward, and is left as an exercise to the reader.

Proposition 6.2.1 *Matrix* A *is TUM if and only if:*

- A^T *is TUM;*

- *matrix* A' *obtained from* A *permuting and/or changing sign to some columns and/or rows is TUM;*

- *matrices* $\begin{bmatrix} \pm 1 & \\ 0 & \\ \vdots & A \\ 0 & \end{bmatrix}$ *and* $\begin{bmatrix} 0 & \\ 0 & \\ \vdots & A \\ 0 & \end{bmatrix}$ *are TUM.*

6.2.1 The Transportation Problem

A typical example of a problem whose "natural" formulation coincides with the "ideal" one is the following *Transportation Problem*.

Consider s origins and t destinations. Each origin $i \in \{1, \ldots, s\}$ has $d_i \geq 0$ units of goods, and each destination $j \in \{1, \ldots, t\}$ needs at least r_j units of the same goods. For each pair $i \in \{1, \ldots, s\}$ and $j \in \{1, \ldots, t\}$, a unit transportation cost (c_{ij}) and a maximum transportation quantity (q_{ij}) are established. We want to determine the transportation policy that minimizes the overall cost.

Indicating with x_{ij} the quantity sent from origin i to destination j, the problem can be formulated as:

$$\min \quad \sum_{i=1}^{s} \sum_{j=1}^{t} c_{ij} x_{ij}$$

$$\sum_{j=1}^{t} x_{ij} \leq d_i \ , \quad i \in \{1, \ldots, s\} \qquad \text{(availability constraints)}$$

$$\sum_{i=1}^{s} x_{ij} \geq r_j \ , \quad j \in \{1, \ldots, t\} \qquad \text{(demand constraints)}$$

$$0 \leq x_{ij} \leq q_{ij} \ , \quad i \in \{1, \ldots, s\} \, , \, j \in \{1, \ldots, t\}. \qquad \text{(capacity constraints)}$$

Changing sign to the availability and capacity constraints, we can convert the problem to its canonical form $\min\{\mathbf{c}^T \mathbf{x} : A\mathbf{x} \geq \mathbf{b}, \mathbf{x} \geq 0\}$, where:

$$
A = \begin{array}{c} \\ 1 \\ \vdots \\ i \\ \vdots \\ s \\ 1 \\ \vdots \\ j \\ \vdots \\ t \\ (1,1) \\ \vdots \\ (i,j) \\ \vdots \\ (s,t) \end{array}
\begin{array}{|ccccc|}
\hline
x_{11} & \cdots & x_{ij} & \cdots & x_{st} \\
\hline
 & & 0 & & \\
 & & \vdots & & \\
 & \cdots & -1 & \cdots & \\
 & & \vdots & & \\
 & & 0 & & \\
 & & 0 & & \\
 & & \vdots & & \\
 & \cdots & +1 & \cdots & \\
 & & \vdots & & \\
 & & 0 & & \\
-1 & & 0 & & 0 \\
 & & \vdots & & \vdots \\
0 & & -1 & & 0 \\
\vdots & \vdots & & \ddots & \vdots \\
0 & \cdots & 0 & \cdots & -1 \\
\hline
\end{array}
\qquad
\mathbf{b} = \begin{bmatrix}
-d_1 \\ \vdots \\ -d_i \\ \vdots \\ -d_s \\ r_1 \\ \vdots \\ r_j \\ \vdots \\ r_t \\ -q_{11} \\ \vdots \\ -q_{ij} \\ \vdots \\ -q_{st}
\end{bmatrix}
\tag{6.1}
$$

Note that each column of matrix A corresponds to a certain variable x_{ij} and has hence exactly 3 elements different from zero, associated with the three inequalities in which x_{ij} appears (the availability constraint for origin i, the demand constraint for destination j, and the capacity constraint on variable x_{ij}). In addition, $a_{ij} \in \{-1, 0, 1\} \ \forall \, i, j$.

It is easy to prove that A is TUM. Indeed, by Proposition 6.2.1 it is possible to eliminate from A the last $s \times t$ rows (these are rows having only one element different from zero), and the total unimodularity of the resulting matrix derives from Theorem 6.2.3, choosing $I_1 = \{1, \ldots, s + t\}$ and $I_2 = \emptyset$. It follows that each basic feasible solution of the linear problem is integer (in the hypothesis that all values of d_i, r_j and q_{ij} are integer), hence a possible integrality constraint on variables x_{ij} would be redundant.

6.3   Cutting Plane Algorithm

In general, the constraint matrix of an ILP problem is not TUM, hence the solution $\mathbf{x}^*$ of its continuous relaxation typically has one or more fractional components. In this case, it is possible to use the so-called *cutting plane* technique.

Consider an ILP problem in standard form $\min\{\mathbf{c}^T\mathbf{x} \ : \ A\mathbf{x} = \mathbf{b} \, , \ \mathbf{x} \geq 0 \ \text{integer}\}$, and let $P := \{\mathbf{x} \geq 0 \ : \ A\mathbf{x} = \mathbf{b}\}$ be the polytope associated with its continuous relaxation.

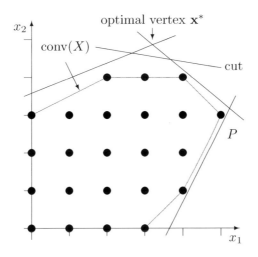

Figure 6.3: Cutting plane

Cutting Plane Algorithm;
    begin
1.   solve the continuous relaxation $\min\{\mathbf{c}^T\mathbf{x} \ : \ A\mathbf{x} = \mathbf{b} \, , \, \mathbf{x} \geq 0\}$,
         and let $\mathbf{x}^*$ be the optimal basic solution found;
2.   **while** $\mathbf{x}^*$ non-integer **do**
         begin
3.   solve the separation problem of $\mathbf{x}^*$ from X, identifying
           an inequality $\alpha^T\mathbf{x} \leq \alpha_0$ valid for X but violated by $\mathbf{x}^*$;
4.   add constraint $\alpha^T\mathbf{x} \leq \alpha_0$ to the current continuous relaxation;
5.   solve the current continuous relaxation (by means of the dual simplex algorithm),
           and let $\mathbf{x}^*$ be the new optimal basic solution identified
         end
    end .

Figure 6.4: Cutting Plane Algorithm

Definition 6.3.1 *Given* $\mathbf{x}^* \in P$ *a cut is an inequality of the type* $\alpha^T\mathbf{x} \leq \alpha_0$ *such that:*

(1) $\alpha^T\mathbf{x} \leq \alpha_0$, $\forall\mathbf{x} \in X := P \cap Z^n$

(2) $\alpha^T\mathbf{x}^* > \alpha_0$.

Conditions (1) and (2) express the fact that inequality $\alpha^T\mathbf{x} \leq \alpha_0$ is valid for X, but violated by point $\mathbf{x}^*$; see Figure 6.3. Given $\mathbf{x}^*$, the problem of identifying a cut of the kind $\alpha^T\mathbf{x} \leq \alpha_0$ is called the *separation problem* of $\mathbf{x}^*$ from X.

The *cutting-plane* procedure illustrated in Figure 6.4 is based on iterative cut generation. The ideal would be to obtain, with little computational effort, deep cuts touching as many integer points as possible. Unfortunately, the two requirements are conflicting and often a compromise has to be found. Note that the convergence of the cutting plane algorithm in general cannot be guaranteed, but it depends on the kind of cuts generated.

6.3.1 Chvátal's Inequalities

Let us now describe a simple procedure to generate valid inequalities for X starting from the continuous relaxation $\min\{\mathbf{c}^T\mathbf{x} \; : \; \mathbf{Ax} = \mathbf{b} \, , \, \mathbf{x} \geq 0\}$.

Consider, as an introductory example, the following ILP problem:

$$
\begin{cases}
\min & -x_1 & -x_2 & -x_3 & & \\
& x_1 & +x_2 & & \leq & 1 \\
& & x_2 & +x_3 & \leq & 1 \\
& x_1 & & +x_3 & \leq & 1 \\
& x_1, & x_2, & x_3 & \geq & 0 \quad \text{integer.}
\end{cases}
$$

It is easy to see that the constraint matrix is not TUM. Converting to standard form, we obtain the system

$$
\begin{cases}
x_1 & +x_2 & & +x_4 & & & = & 1 \\
& x_2 & +x_3 & & +x_5 & & = & 1 \\
x_1 & & +x_3 & & & +x_6 & = & 1.
\end{cases}
$$

The optimal solution of the continuous relaxation is $\mathbf{x}^* = \left[\frac{1}{2}, \frac{1}{2}, \frac{1}{2}, 0, 0, 0\right]^T$ hence $z_{PL} = -\frac{3}{2}$. The optimal value z_{PLI} can instead be identified by inspection, and is equal to $z_{PLI} = -1$.

If we multiply by $\frac{1}{2}$ the three equations and add them together, we obtain the equation:

$$
x_1 + x_2 + x_3 + \frac{1}{2}x_4 + \frac{1}{2}x_5 + \frac{1}{2}x_6 = \frac{3}{2}.
$$

By construction this equation is valid for P, hence it is also valid for $X \subset P$. Reducing the coefficients of variables x_j until they become integer, we obtain the following inequality:

$$
x_1 + x_2 + x_3 + 0x_4 + 0x_5 + 0x_6 \leq \frac{3}{2}.
$$

This inequality is valid for P, given that $\mathbf{x} \in P$ implies $\mathbf{x} \geq 0$, and thus it is also valid for X. We can now exploit the integrality of the variables: for all *integer* $\mathbf{x}$, the left-hand side must be integer, and hence it is possible to decrease the right-hand side from $\frac{3}{2}$ to the nearest integer value. In this way, we obtain the valid inequality for X (but not necessarily for P):

$$
x_1 + x_2 + x_3 \leq 1.
$$

Note that this inequality cuts the fractional point $\mathbf{x}^* = \left[\frac{1}{2}, \frac{1}{2}, \frac{1}{2}, 0, 0, 0\right]^T$.

In general, the Chvátal procedure for the generation of valid inequalities for X operates as follows. For all $r \in \Re$ let $\lfloor r \rfloor$ be the truncation of r to the nearest lower integer, i.e.

$$\lfloor r \rfloor := \max\{i \in Z : i \le r\}.$$

For example, $\lfloor 2 \rfloor = 2$, $\lfloor 3.9 \rfloor = 3$, $\lfloor -5.1 \rfloor = -6$. Given a vector $\mathbf{v} = [v_1, \ldots, v_t]^T \in \Re^t$ let $\lfloor \mathbf{v} \rfloor := [\lfloor v_1 \rfloor, \ldots, \lfloor v_t \rfloor]^T$.

Given system $A\mathbf{x} = \mathbf{b}$, $\mathbf{x} \ge 0$, choose randomly $\mathbf{u} \in \Re^m$ and obtain by linear combination the valid equation for P:

$$\mathbf{u}^T A\mathbf{x} = \mathbf{u}^T \mathbf{b}.$$

By means of truncation, we can obtain the "weakened" inequality

$$\alpha^T \mathbf{x} := \lfloor \mathbf{u}^T A \rfloor \mathbf{x} \le \mathbf{u}^T \mathbf{b},$$

where coefficients α_j are defined as $\alpha_j := \lfloor \mathbf{u}^T A_j \rfloor := \lfloor \sum_{i=1}^m u_i a_{ij} \rfloor$ for all $j = 1, \ldots, n$.

At last, defining $\alpha_0 := \lfloor \mathbf{u}^T \mathbf{b} \rfloor$ we obtain the valid inequality for X (but not necessarily for P)

$$\alpha^T \mathbf{x} \le \alpha_0.$$

It is possible to prove that, given a fractional vertex $\mathbf{x}^*$ of P, there always exists a vector $\mathbf{u} \in \Re^m$ for which the corresponding inequality $\alpha^T \mathbf{x} \le \alpha_0$ is violated by $\mathbf{x}^*$. With the Chvátal procedure it is therefore always possible to cut any fractional vertex of P, by choosing vector $\mathbf{u}$ appropriately.

However, it would be wrong to think that all valid inequalities for X are obtained *directly* starting from the initial system $A\mathbf{x} = \mathbf{b}$: there exist valid inequalities for X that are not obtained as $\lfloor \mathbf{u}^T A \rfloor \mathbf{x} \le \lfloor \mathbf{u}^T \mathbf{b} \rfloor$ for any $\mathbf{u} \in \Re^m$.

Indicating with $A^{(1)}\mathbf{x} \le \mathbf{b}^{(1)}$ the family of inequalities $\lfloor \mathbf{u}^T A \rfloor \mathbf{x} \le \lfloor \mathbf{u}^T \mathbf{b} \rfloor$, $\mathbf{u} \in \Re^m$, we can define the *first Chvátal's closure of P* as

$$P_1 := \{\mathbf{x} \ge 0 : A\mathbf{x} = \mathbf{b}, A^{(1)}\mathbf{x} \le \mathbf{b}^{(1)}\} \subseteq P,$$

where $P_1 = P$ if and only if P does not have fractional vertices. Although system $A^{(1)}\mathbf{x} \le \mathbf{b}^{(1)}$, in principle, contains ∞^m inequalities, it is possible to prove that only a finite number of its inequalities is necessary to describe P_1, hence P_1 is a polytope.

The addition of new constraints $A^{(1)}\mathbf{x} \le \mathbf{b}^{(1)}$ allows one to improve the initial formulation $A\mathbf{x} = \mathbf{b}$, $\mathbf{x} \ge 0$. If $\text{conv}(X) = P_1$, this formulation is ideal. Otherwise there exist fractional vertices of P_1, that can be cut generating new Chvátal's inequalities obtained combining those defining P_1. In this way we obtain a new family of inequalities, say $A^{(2)}\mathbf{x} \le \mathbf{b}^{(2)}$, which theoretically contains all constraints of the kind

$$\left\lfloor \mathbf{u}^T A + \mathbf{w}^T A^{(1)} \right\rfloor \mathbf{x} \le \left\lfloor \mathbf{u}^T \mathbf{b} + \mathbf{w}^T \mathbf{b}^{(1)} \right\rfloor, \quad \mathbf{u} \in \Re^m, \ \mathbf{w} \ge 0;$$

note that $\mathbf{w} \ge 0$, given that constraints $A^{(1)}\mathbf{x} \le \mathbf{b}^{(1)}$ are inequalities.

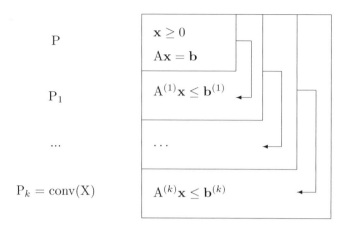

Figure 6.5: Chvátal's closure

Adding these inequalities to the current system $A\mathbf{x} = \mathbf{b}$, $A^{(1)}\mathbf{x} \leq \mathbf{b}^{(1)}$, $\mathbf{x} \geq 0$, we obtain the *second Chvátal's closure*, i.e., polytope

$$P_2 := \{\mathbf{x} \geq 0 \; : \; A\mathbf{x} = \mathbf{b}\,,\; A^{(1)}\mathbf{x} \leq \mathbf{b}^{(1)}\,,\; A^{(2)}\mathbf{x} \leq \mathbf{b}^{(2)}\} \subseteq P_1.$$

By iterating the procedure for a finite number of times, say k times, we obtain a succession of polytopes $P_1, P_2, \ldots, P_k$ such that (see Figure 6.5)

$$P \supset P_1 \supset P_2 \supset \ldots \supset P_k \; = \; \mathrm{conv}(X).$$

Value k is said to be the *Chvátal's rank* of $\mathrm{conv}(X)$ with respect to P. Polytopes without fractional vertices have rank 0 (initial formulation = ideal formulation), they have rank 1 when all inequalities defining $\mathrm{conv}(X)$ are directly obtained from the initial formulation, and they have rank $k \geq 2$ in the other cases. Note that rank k, even if finite, can be very large: there exist cases in which k is even greater than the number of variables of the problem.

The procedure described makes it possible to obtain an ideal formulation of $\mathrm{conv}(X)$ starting from any initial formulation. The number of necessary inequalities, even though finite, is generally huge: already in system $A^{(1)}\mathbf{x} = \mathbf{b}^{(1)}$ we can count approximately $\binom{n}{m}$ essential inequalities! On the other hand, in the *cutting-plane* algorithm, an ideal formulation of the *entire* polytope $\mathrm{conv}(X)$ is not required: what we want to identify is a set of cuts that "brings out" an integer solution, highlighting it as the optimal vertex (for the given objective function) of the current continuous relaxation. To that end, it is possible to use a method, called the *Gomory's cutting plane algorithm*, that generates only some Chvátal's inequalities, chosen from those violated by the current fractional vertex $\mathbf{x}^*$.

6.3.2 Gomory's Cuts

The idea is to generate a cut using the information associated with the basis B corresponding to the optimal basic solution $\mathbf{x}^*$ of the current continuous relaxation. If $\mathbf{x}^*$ is integer, it is not necessary to generate any cut. Otherwise, there exists at least one fractional component x_h^*. Variable x_h must then be a basic variable on a certain row t (say) of the optimal tableau associated with $\mathbf{x}^*$ (the one for which $\beta[t] = h$).

The equation associated with row t of the optimal tableau is the following:

$$x_h + \sum_{j \in F} \bar{a}_{tj} x_j = \bar{b}_t \, (= x_h^*), \tag{6.2}$$

where F is the set of indices of non-basic variables. Row t is said to be the *generating row* for the cut. Equation (6.2) is, by construction, valid for P (it is an equation of the form $\mathbf{u}^T A \mathbf{x} = \mathbf{u}^T \mathbf{b}$, where $\mathbf{u}^T$ is the t-th row of the current matrix B^{-1}). By applying Chvátal's truncation procedure starting from this equation, we immediately obtain *Gomory's cut*:

$$x_h + \sum_{j \in F} \lfloor \bar{a}_{tj} \rfloor x_j \leq \lfloor \bar{b}_t \rfloor. \tag{6.3}$$

By construction, this inequality is valid for X, but violated by $\mathbf{x}^*$. Indeed, given that $x_j^* = 0$ for all $j \in F$ and that $\bar{b}_t = x_h^*$ is fractional, we have:

$$\left(x_h + \sum_{j \in F} \lfloor \bar{a}_{tj} \rfloor x_j \right)_{\mathbf{x}=\mathbf{x}^*} = x_h^* = \bar{b}_t > \lfloor \bar{b}_t \rfloor.$$

Formulation (6.3) is said to be the *integer form* of the cut. There exists an equivalent formulation, called *fractional form*, which is obtained by subtracting (6.3) from (6.2):

$$\sum_{j \in F} (\bar{a}_{tj} - \lfloor \bar{a}_{tj} \rfloor) x_j \geq \bar{b}_t - \lfloor \bar{b}_t \rfloor,$$

i.e.

$$\sum_{j \in F} \varphi(\bar{a}_{tj}) x_j \geq \varphi(\bar{b}_t), \tag{6.4}$$

where for all $r \in \Re$ we indicate with

$$\varphi(r) := r - \lfloor r \rfloor \geq 0$$

the *fractional part* of r. For example, $\varphi(5) = 0$, $\varphi(\frac{8}{3}) = \frac{2}{3}$, $\varphi(-\frac{8}{3}) = 1 - \varphi(\frac{8}{3}) = \frac{1}{3}$.

Note that the left-hand side of (6.4) is zero for $\mathbf{x} = \mathbf{x}^*$, while the right-hand side is strictly positive. In addition, all coefficients in (6.4) are greater than or equal to zero.

Converting (6.3) to standard form, we obtain $x_h + \sum_{j \in F} \lfloor \bar{a}_{tj} \rfloor x_j + s = \lfloor \bar{b}_t \rfloor$, where the slack variable $s \geq 0$ will be integer for all integer $\mathbf{x}$, given that all coefficients are integer. Subtracting equation (6.2), we obtain the constraint in standard form:

$$\sum_{j \in F} -\varphi(\bar{a}_{tj}) x_j + s = -\varphi(\bar{b}_t) \, , \ s \geq 0 \text{ integer.} \tag{6.5}$$

This constraint can easily be added to the current tableau maintaining the dual feasibility required for the application of the dual simplex method at step 5 of the *cutting plane* algorithm.

		x_h					$x_j \, , \, j \in F$		s
$-z_{PL}$	0	...	0	...	0	...	$\bar{c}_j \geq 0$	...	0
									0
generating row t $\quad \bar{b}_t$	0	...	1	...	0	...	$\bar{a}_{tj}$	...	$\vdots$
									0
Gomory's cut $\quad -\varphi(\bar{b}_t)$	0	...	0	...	0	...	$-\varphi(\bar{a}_{tj})$	...	1

Example 1

Consider the following ILP problem:

$$\begin{cases} \min & -x_2 \\ & 3x_1 & +2x_2 & \leq & 6 \\ & -3x_1 & +2x_2 & \leq & 0 \\ & x_1, & x_2 & \geq & 0 \text{ integer.} \end{cases}$$

By applying the primal simplex algorithm to the continuous relaxation we obtain the tableaux:

		x_1	x_2	x_3	x_4
$-z$	0	0	-1	0	0
x_3	6	3	2	1	0
x_4	0	-3	②	0	1

$x_3 = 6 - 3x_1 - 2x_2$

$x_4 = 3x_1 - 2x_2$

		x_1	x_2	x_3	x_4
$-z$	0	$-\frac{3}{2}$	0	0	$\frac{1}{2}$
x_3	6	⑥	0	1	-1
x_2	0	$-\frac{3}{2}$	1	0	$\frac{1}{2}$

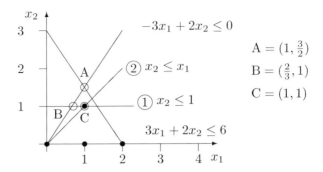

Figure 6.6: Graphical representation of Example 1

		x_1	x_2	x_3	x_4
$-z$	$\frac{3}{2}$	0	0	$\frac{1}{4}$	$\frac{1}{4}$
x_1	1	1	0	$\frac{1}{6}$	$-\frac{1}{6}$
x_2	$\frac{3}{2}$	0	1	$\frac{1}{4}$	$\frac{1}{4}$

The optimal solution $\mathbf{x}^* = \left[1, \frac{3}{2}, 0, 0\right]^T$ has value $z_{PL} = -\frac{3}{2}$ (vertex A in Figure 6.6). Let us generate a Gomory's cut starting from the generating row $t = 2$

$$x_2 + \tfrac{1}{4}x_3 + \tfrac{1}{4}x_4 = \tfrac{3}{2} \;\Rightarrow\; x_2 + 0x_3 + 0x_4 \le \left\lfloor \tfrac{3}{2} \right\rfloor,$$

obtaining constraint $x_2 \le 1$ (cut 1 in Figure 6.6). By subtracting the two constraints and adding the integer slack variable $x_5 \ge 0$, we obtain the fractional form:

$$\tfrac{1}{4}x_3 + \tfrac{1}{4}x_4 \ge \tfrac{1}{2} \;\Rightarrow\; -\tfrac{1}{4}x_3 - \tfrac{1}{4}x_4 + x_5 = -\tfrac{1}{2}.$$

We can now create the augmented tableau:

	x_1	x_2	x_3	x_4	x_5
$\frac{3}{2}$	0	0	$\frac{1}{4}$	$\frac{1}{4}$	0
1	1	0	$\frac{1}{6}$	$-\frac{1}{6}$	0
$\frac{3}{2}$	0	1	$\frac{1}{4}$	$\frac{1}{4}$	0
$-\frac{1}{2}$	0	0	$\boxed{-\frac{1}{4}}$	$-\frac{1}{4}$	1

$$
\begin{aligned}
x_5 &= -\tfrac{1}{2} + \tfrac{1}{4}x_3 + \tfrac{1}{4}x_4 = \\
&\quad -\tfrac{1}{2} + \tfrac{1}{4}(6 - 3x_1 - 2x_2) + \\
&\quad\quad +\tfrac{1}{4}(3x_1 - 2x_2) = \\
&= 1 - x_2.
\end{aligned}
$$

Note that the new slack variable x_5 can be obtained as a function of the single variables x_1 and x_2, by means of the simple substitutions shown near the previous tableau. This allows us to represent the cuts obtained in the space of the only original variables of the problem.

By applying the dual simplex, we obtain the new optimal tableau:

		x_1	x_2	x_3	x_4	x_5
$-z$	1	0	0	0	0	1
x_1	$\frac{2}{3}$	1	0	0	$-\frac{1}{3}$	$\frac{2}{3}$
x_2	1	0	1	0	0	1
x_3	2	0	0	1	1	-4

Solution $\mathbf{x}^* = \left[\frac{2}{3}, 1, 2, 0, 0\right]^T$ is still fractional (vertex B in Figure 6.6). By generating a Gomory's cut from row $t = 1$, we obtain constraint $x_1 - x_4 \leq \left\lfloor \frac{2}{3} \right\rfloor = 0$, from which, by substituting $x_4 = 3x_1 - 2x_2$, we obtain constraint $-2x_1 + 2x_2 \leq 0$ (cut 2 in Figure 6.6). Working directly on the last tableau we easily obtain the augmented tableau:

		x_1	x_2	x_3	x_4	x_5	x_6
$-z$	1	0	0	0	0	1	0
x_1	$\frac{2}{3}$	1	0	0	$-\frac{1}{3}$	$\frac{2}{3}$	0
x_2	1	0	1	0	0	1	0
x_3	2	0	0	1	1	-4	0
x_6	$-\frac{2}{3}$	0	0	0	$\left(-\frac{2}{3}\right)$	$-\frac{2}{3}$	1

$$x_6 = -\frac{2}{3} + \frac{2}{3}x_4 + \frac{2}{3}x_5 =$$
$$= -\frac{2}{3}(3x_1 - 2x_2) +$$
$$+\frac{2}{3}(1 - x_2) =$$
$$= 2x_1 - 2x_2$$

where the integer $x_6 \geq 0$ is the new slack variable, which can be obtained as a function of x_1 and x_2 as slack of $-2x_1 + 2x_2 \leq 0$, or directly from the augmented tableau by means of successive substitutions. By applying the dual simplex, we obtain the optimal tableau

		x_1	x_2	x_3	x_4	x_5	x_6
$-z$	1	0	0	0	0	1	0
x_1	1	1	0	0	0	1	$-\frac{1}{2}$
x_2	1	0	1	0	0	1	0
x_3	1	0	0	1	0	-5	$\frac{3}{2}$
x_4	1	0	0	0	1	1	$-\frac{3}{2}$

The optimal solution of the continuous relaxation is now $\mathbf{x}^* = [1, 1, 1, 1, 0, 0]^T$ and corresponds to the integer vertex C of Figure 6.6. Note that the current formulation is not ideal: the polytope of Figure 6.6 still has a fractional vertex, given that the description of conv(X) would require the additional constraint $x_1 + x_2 \leq 2$. However, for the objective function used, this latter constraint is not necessary.

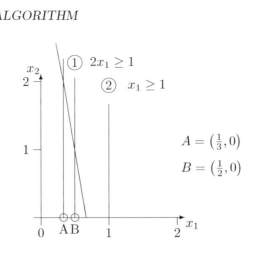

Figure 6.7: Graphic representation of example 2

Example 2

Consider the following ILP problem:

$$\begin{cases} \min & x_1 & +x_2 & & \\ & 6x_1 & +x_2 & \leq & 4 \\ & 3x_1 & & \geq & 1 \\ & x_1, & x_2 & \geq & 0 \text{ integer.} \end{cases}$$

The formulation above is clearly not tight: by exploiting the integrality constraint, the second constraint can be rewritten as $x_1 \geq 1$, which proves that no integer feasible solution exists, given that the first constraint is necessarily violated if $x_1 \geq 1$. However, we apply the procedure starting from the given formulation. The initial tableau is:

		x_1	x_2	x_3	x_4
$-z$	0	1	1	0	0
x_3	4	6	1	1	0
x_4	-1	-3	0	0	1

$$x_3 = 4 - 6x_1 - x_2$$
$$x_4 = -1 + 3x_1$$

By applying the dual simplex algorithm, we immediately arrive to the optimal tableau:

		x_1	x_2	x_3	x_4
$-z$	$-\frac{1}{3}$	0	1	0	$\frac{1}{3}$
x_3	2	0	1	1	2
x_1	$\frac{1}{3}$	1	0	0	$-\frac{1}{3}$

Solution $\mathbf{x}^* = \left[\frac{1}{3}, 0, 2, 0\right]^T$ corresponds to vertex A of Figure 6.7. Let us generate from row $t = 2$ the cut $x_1 - x_4 \leq 0$, i.e., $x_1 - (-1 + 3x_1) = 1 - 2x_1 \leq 0$, with slack variable $x_5 = 2x_1 - 1$. The augmented tableau is:

	x_1	x_2	x_3	x_4	x_5
$-z$	$-\frac{1}{3}$	0	1	0	$\frac{1}{3}$	0
x_3	2	0	1	1	2	0
x_1	$\frac{1}{3}$	1	0	0	$-\frac{1}{3}$	0
x_5	$-\frac{1}{3}$	0	0	0	$\left(-\frac{2}{3}\right)$	1

$x_5 = -\frac{1}{3} + \frac{2}{3}x_4 =$
$= -\frac{1}{3} + \frac{2}{3}(-1 + 3x_1) =$
$= -1 + 2x_1$

Re-optimizing with the dual simplex, we obtain the optimal tableau

	x_1	x_2	x_3	x_4	x_5
$-z$	$-\frac{1}{2}$	0	1	0	0	$\frac{1}{2}$
x_3	1	0	1	1	0	3
x_1	$\frac{1}{2}$	1	0	0	0	$-\frac{1}{2}$
x_4	$\frac{1}{2}$	0	0	0	1	$-\frac{3}{2}$

Generating, for instance, a Gomory's cut from row $t = 2$, we obtain cut $x_1 - x_5 \leq 0$, i.e, $x_1 - (-1 + 2x_2) = 1 - x_1 \leq 0$, and the corresponding slack variable $x_6 = x_1 - 1$. By creating the augmented tableau, we obtain:

	x_1	x_2	x_3	x_4	x_5	x_6
$-z$	$-\frac{1}{2}$	0	1	0	0	$\frac{1}{2}$	0
x_3	1	0	1	1	0	3	0
x_1	$\frac{1}{2}$	1	0	0	0	$-\frac{1}{2}$	0
x_4	$\frac{1}{2}$	0	0	0	1	$-\frac{3}{2}$	0
x_6	$-\frac{1}{2}$	0	0	0	0	$\left(-\frac{1}{2}\right)$	1

$x_6 = -\frac{1}{2} + \frac{1}{2}x_5 =$
$= -\frac{1}{2} + \frac{1}{2}(-1 + 2x_2) =$
$= -1 + x_1$

With a pivot operation, we obtain the tableau

	x_1	x_2	x_3	x_4	x_5	x_6
$-z$	-1	0	1	0	0	0	1
x_3	-2	0	1	1	0	0	6
x_1	1	1	0	0	0	0	-1
x_4	2	0	0	0	1	0	-3
x_5	1	0	0	0	0	1	-2

The equation associated with row 1 is inconsistent with the constraint $\mathbf{x} \geq 0$. Therefore, the current continuous relaxation is impossible, i.e., $P = \emptyset$, which implies $X = \emptyset$ given that $X \subseteq P$.

6.3.3 Computational considerations on Gomory's cuts

The generation of Gomory's cuts requires little computational effort and is a general procedure particularly simple and elegant. A fundamental theoretical property of Gomory's cuts is that their use can ensure the convergence of the cutting plane algorithm in a finite (although sometimes very large) number of iterations. This is not necessarily true using other techniques for the generation of cuts. For instance, it is easy to prove that the current fractional solution $\mathbf{x}^*$ can always be separated by means of the valid cut $\sum_{j \in F} x_j \geq 1$ where, as usual, F is the set of indices of non-basic variables. By using these cuts, however, convergence after a finite number of iterations is not guaranteed, and in some cases the integer optimal solution can only be reached asymptotically.

Despite its undisputed theoretical importance, Gomory's method is generally considered to be ineffective because the quality of the cuts generated tends to iteratively decrease: after few iterations the cuts are not too deep and therefore not very effective. However, recent studies have re-evaluated this approach, showing that it can be integrated with other techniques in order to be effective with some classes of ILP problems.

An important point of the method is the choice of the t row from which to generate the cut. Note that, when the cost vector $\mathbf{c}$ is integer, the generating row could even coincide with row 0 (objective function) of the tableau. In this case, the corresponding equation is $-z + \sum_{j \in F} \bar{c}_j x_j = \bar{c}_0$ (remember that the column associated with "basic variable" z has been eliminated from the tableau). If $\bar{c}_0$ is not integer, from this equation we can obtain cut $\sum_{j \in F} \varphi(\bar{c}_j) x_j \geq \varphi(\bar{c}_0)$, which can be added without problems to the current tableau given that variable z does not explicitly appear.

A rule commonly used for the choice of the generating row t is that of preferring the row with maximum $\varphi(\bar{b}_t)$. In this way, we choose the constraint that is the "most violated" by $\mathbf{x}^*$, and hence most likely the "deepest" constraint. However, since nothing prohibits to add several cuts at the same time, a second possibility is to generate, when possible, *all* Gomory's cuts with violation $\varphi(\bar{b}_t) > \varepsilon$, where ε is a properly defined parameter (for instance, $\varepsilon = 0.01$). In this way, the re-optimization of the tableau with the dual simplex usually requires a greater computational effort, but convergence typically occurs after much fewer iterations given that many deep cuts generated from the first tableaux have been exploited.

6.4 Branch-and-Bound

An alternative algorithm for solving an ILP problem (and, more generally, a combinatorial optimization problem) is known in the literature as *implicit enumeration*, or *branch-and-bound*. It is a *divide and conquer* procedure that (recursively) casts the solution of a "difficult" ILP problem into the solution of two "easier" ILP subproblems.

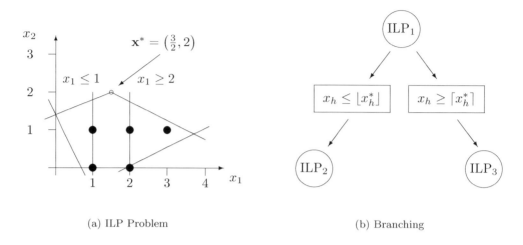

(a) ILP Problem   (b) Branching

Figure 6.8: Branch-and-Bound

Consider the example shown in Figure 6.8(a). By solving the continuous relaxation, we identify the optimal solution $\mathbf{x}^*$ with fractional $x_1^* = \frac{3}{2}$. Clearly, the optimal integer solution will have either $x_1 \leq 1$ or $x_1 \geq 2$. However, the logic inequality "$x_1 \leq 1$ or $x_1 \geq 2$" cannot be expressed as a linear constraint, and hence it is not a valid cut.

However, we can proceed as follows. Start by imposing the single condition $x_1 \leq 1$, and then solve the corresponding ILP problem by identifying the optimal integer solution $\mathbf{x}'_{OPT}$. Consider then the single condition $x_1 \geq 2$, and solve the corresponding ILP problem by identifying the optimal integer solution $\mathbf{x}''_{OPT}$. At this stage, it is immediate to find the optimal solution $\mathbf{x}_{OPT}$ of the original problem, choosing between $\mathbf{x}'_{OPT}$ and $\mathbf{x}''_{OPT}$ the solution with minimum cost.

In general, let $\min\{\mathbf{c}^T\mathbf{x} : A\mathbf{x} = \mathbf{b}, \mathbf{x} \geq 0 \text{ integer}\}$ be the original ILP problem (problem ILP_1). Solve the continuous relaxation $\min\{\mathbf{c}^T\mathbf{x} : A\mathbf{x} = \mathbf{b}, \mathbf{x} \geq 0\}$, and identify an optimal basic solution $\mathbf{x}^*$. If $\mathbf{x}^*$ is integer, no further action is necessary. Otherwise, choose a fractional variable x_h^* and create two subproblems

$$ILP_2 : \min\{\mathbf{c}^T\mathbf{x} : A\mathbf{x} = \mathbf{b}, x_h \leq \lfloor x_h^* \rfloor, \mathbf{x} \geq 0 \text{ integer}\}$$
$$ILP_3 : \min\{\mathbf{c}^T\mathbf{x} : A\mathbf{x} = \mathbf{b}, x_h \geq \lceil x_h^* \rceil, \mathbf{x} \geq 0 \text{ integer}\},$$

where

$$\lceil r \rceil := \min\{i \in Z : i \geq r\}$$

is the rounding up (ceiling) of a given value $r \in \Re$.

This operation is called *branching*, and can be graphically represented as in Figure 6.8(b). Variable x_h is called the *branching variable*. Problems ILP_2 and ILP_3 are called the *children* of ILP_1: they are obtained from ILP_1 by adding a branching constraint.

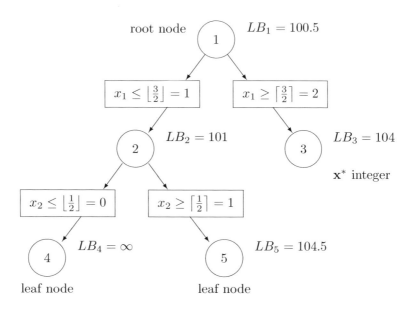

Figure 6.9: Branching tree

Now we solve ILP$_2$ and ILP$_3$, in sequence, by recursively reapplying the procedure: for each of the two subproblems, we solve the continuous relaxation and if necessary we perform again the branching operation. In this way, we obtain a hierarchical sequence of subproblems, that become more and more constrained, and hence easier to solve.

Figure 6.9 shows the *branching tree* typically used to describe the procedure. Node 1 corresponds to the original ILP problem (*root* node). Each child node is obtained from the parent node by adding the constraint shown on the branch connecting them. In general, the constraints characterizing the problem associated with a given node t are:

(1) the original constraints $\mathbf{A}\mathbf{x} = \mathbf{b}$, $\mathbf{x} \geq 0$ integer, and

(2) all the *branching constraints* found in the branching tree along the path from the root node to node t.

Consider now the convergence of the method. Since, by hypothesis, $P = \{\mathbf{x} \geq 0 : \mathbf{A}\mathbf{x} = \mathbf{b}\}$ is bounded, the total number of nodes of the branching tree is finite. In the worst case, however, this number can be very large. For instance, consider the easiest case in which $0 \leq x_j \leq 1$, x_j integer, for all $j \in \{1, \ldots, n\}$. At each branching operation, we will then have $x_h \leq \lfloor x_h^* \rfloor = 0$ or $x_h \geq \lceil x_h^* \rceil = 1$, hence in both cases variable x_h will be fixed (to either 0 or 1). As we go down along the branching tree, the number of fixed variables increases and thus, in the worst case, the tree will have a depth of n and will contain 2^n leaf nodes (plus $2^n - 1$ intermediate nodes).

6.4.1   Fathoming criteria (pruning)

In practice, there is no need to *explicitly* consider all nodes potentially present in the branching tree. Indeed, some leaf nodes of the current branching tree may correspond to problems whose continuous relaxation is infeasible. This happens when the initial constraints $A\mathbf{x} = \mathbf{b}$, $\mathbf{x} \geq 0$, are incompatible with the branching constraints related to the path that from the root node leads to the leaf node in question. These problems do not generate children.

The same happens when the continuous relaxation PL_t of the current node t has optimal integer solution $\mathbf{x}^*_{PL_t}$. Let

- $\mathbf{x}_{OPT}$ = *current optimal solution* or *incumbent* (the best integer solution found so far)

- $z_{OPT} = \mathbf{c}^T \mathbf{x}_{OPT}$ = incumbent cost.

If $\mathbf{c}^T \mathbf{x}^*_{PL_t} < z_{OPT}$, we update $z_{OPT} := \mathbf{c}^T \mathbf{x}^*_{PL_t}$ and $\mathbf{x}_{OPT} := \mathbf{x}^*_{PL_t}$; in any case, there is no need to further analyze the current node t, and we can consider other nodes of the branching tree.

There exists a third criterion, called *bounding*, which often allows one to eliminate a large number of nodes of the branching tree. Suppose to have solved the continuous relaxation PL_t of the problem associated with the current node t, and let $\mathbf{x}^*_{PL_t}$ be the optimal solution identified. If $\mathbf{x}^*_{PL_t}$ is fractional, we will have to continue generating the children of node t, hoping that they enable us to find an integer solution *better* than the incumbent $\mathbf{x}_{OPT}$. Before proceeding, however, we can note that value $LB_t := \mathbf{c}^T \mathbf{x}^*_{PL_t}$ is an optimistic estimate (*lower bound*) of the value of the best integer solution that can be obtained starting from node t. If $LB_t \geq z_{OPT}$, this solution will not be better than the incumbent, hence there is no need to further process node t.

In the example of Figure 6.9, the values LB_t are shown in correspondence of the nodes of the tree. Note that the values LB_t cannot decrease going from parent to child, since they correspond to problems becoming more and more constrained. The current leaf nodes are: node 3, the integer solution of which is $\mathbf{x}^*_{PL_3}$ with value 104 (hence $z_{OPT} = 104$ in this moment); node 4 with $LB_4 = +\infty$ (problem PL_4 infeasible); and node 5, the fractional solution of which is $\mathbf{x}^*_{PL_5}$ with value $LB_5 = 104.5$. The algorithm would now generate the two children of node 5. However, given that $LB_5 \geq z_{OPT}$, any further processing of node 5 is useless.

Summarizing, there are three *fathoming (or pruning) criteria* to declare the current node t *fathomed/pruned*, and hence to avoid considering its children:

(1)   solution $\mathbf{x}^*_{PL_t}$ of the continuous relaxation PL_t is integer;

(2)   the continuous relaxation PL_t is infeasible;

(3) $LB_t \geq z_{OPT}$, where $LB_t := \mathbf{c}^T \mathbf{x}^*_{PL_t}$ and z_{OPT} is the cost of the current optimal integer solution (*bounding* criterion).

Note that, if $\mathbf{c}$ is integer, then z_{OPT} will be integer as well. In this case, it is then possible to define $LB_t := \left\lceil \mathbf{c}^T \mathbf{x}^*_{PL_t} \right\rceil$, which can make the bounding criterion more effective.

The use of the bounding criterion cannot guarantee an improvement of the complexity in the *worst case* of the procedure. In practice, however, it can often allow the elimination of most of the nodes of the branching tree. Clearly this is all the more true, the tighter the initial formulation of the problem is: the lower bounds are more accurate and the fathoming criteria more effective.

6.4.2 Implementation of the Algorithm

An implementation of the *branch-and-bound* algorithm must first and foremost establish the rule to choose the branching variable x^*_h, rule that has a very significant impact on the total number of nodes generated. A simple option is to choose the variable whose fractional part is as near as possible to 0.5, such that the branching condition significantly acts on both child nodes. The effectiveness of such rule is however debatable: recent computation studies have indeed shown that the rule does not give significantly different results from those obtained by choosing the branching variable completely at random (among the fractional ones). In practice, the rules to prefer are those estimating the increase of lower bound obtainable at the two child nodes, preferring the variable in which the product between the increases is maximum.

We will also have to establish the rule with which to select the node to process in the current iteration. There are two main techniques for visiting the branching tree:

Depth First Search (Deepest Nodes First)

The deepest not yet processed node is systematically chosen, simulating a recursive procedure: given a parent node, consider immediately the first of its two children, until one of the fathoming conditions forces the algorithm to backtrack.

Pros: the implementation is typically easier, since it can exploit the recursion of the programming language used; LPs are typically easier to solve.

Cons: in case of "wrong" choices, backtracking occurs only after having explored the *whole* wrong subtree.

Best-bound First (Most Promising Nodes First)

The choice of the node to be processed is made by choosing the not yet processed node t with minimum LB_t, and hence likely closer to the optimal solution.

Pros: typically less nodes are generated.

Cons: the processing tends to remain at shallow levels of the branching tree, and therefore the problems are less constrained and hardly lead to an update of the incumbent.

This latter disadvantage can be minimized by identifying, in a heuristic way, a good
integer solution to be used to initialize the incumbent.

Formalization of the Algorithm

Figure 6.10 shows a possible implementation of the *branch-and-bound* algorithm, in which
each subproblem PL_t is first solved and then inserted (with its LB_t value) into the queue
of the open problems waiting to be processed. The data structure used is the following:

- m = last node of the branching tree;

- $\mathbf{x}_{OPT}$ = current optimal integer solution (incumbent);

- $z_{OPT} = \mathbf{c}^T \mathbf{x}_{OPT}$ = incumbent cost;

- Q = queue of the *active* open nodes (those that can still generate child nodes).

For each node t of the current branching tree, let:

- $parent[t] = \pm p$, where p is the parent node of t in the branching tree (positive for
  the left child, negative for the right child);

- $LB[t]$ = lower bound associated with t ;

- $vbranch[t]$ = index h of the branching variable x_h;

- $value[t]$ = value x_h^* of the branching variable.

For the sake of simplicity, we assume that the optimal solution of an infeasible LP problem
is a dummy vector $\mathbf{x}^*$ with $\mathbf{c}^T \mathbf{x}^* = +\infty$.

Steps 1-5 process the root node. At step 7, we choose the node t to be processed (by
depth first or best-bound first search). At step 9 we generate the two children of the
current node t, each of which is solved (step 12) and inserted in the queue Q of the
active nodes, together with the corresponding lower bound and branching variable (step
15). The update of the current optimal solution $\mathbf{x}_{OPT}$ occurs at step 14; note that the
new value z_{OPT} is compared with all values $LB[j]$ of nodes $j \in Q$, in the hope that the
bounding criterion allows the elimination of some nodes from Q. At steps 5 and 15, when
$LB[m] < z_{OPT}$ we have that $\mathbf{x}^*$ must necessarily be fractional: if $\mathbf{x}^*$ were integer, after
the update of $\mathbf{x}_{OPT}$ we would necessarily have $LB[m] \geq z_{OPT}$.

As already said, if $\mathbf{c}$ is integer, the lower bound $LB[m]$ can be computed as $\lceil \mathbf{c}^T \mathbf{x}^* \rceil$ at
steps 3 and 13.

Branch-and-Bound Algorithm ;
  begin
1. $m := 1$; $parent[1] := 0$; $Q := \emptyset$;
     $z_{OPT} :=$ heuristic solution value (possibly $+\infty$);
2.   solve the continuous relaxation $\min\{\mathbf{c}^T\mathbf{x} \; : \; A\mathbf{x} = \mathbf{b} \, , \, \mathbf{x} \geq 0\}$, and let
     $\mathbf{x}^*$ be the optimal solution found;
3. $LB[1] := \mathbf{c}^T\mathbf{x}^*$;
4.   **if** $(\mathbf{x}^*$ integer$)$ **and** $(\mathbf{c}^T\mathbf{x}^* < z_{OPT})$ **then**
          begin
              $\mathbf{x}_{OPT} = \mathbf{x}^*$; $z_{OPT} := \mathbf{c}^T\mathbf{x}^*$
          end ;
5.   **if** $LB[1] < z_{OPT}$ **then**
          begin
              choose the fractional branching variable x_h^* ;
              $vbranch[1] := h$; $value[1] := x_h^*$;
              $Q := \{1\}$
          end ;
6.   **while** $Q \neq \emptyset$ **do** /* process the active open nodes*/
          begin
7.   choose a node $t \in Q$, and set $Q := Q \setminus \{t\}$;
8. $h := vbranch[t]$; $val := value[t]$;
9.   **for** $child := 1$ **to** 2 **do** /* generate the children of node t */
              begin
10. $m := m + 1$;
                  if $child = 1$
                      then $parent[m] := t$
                      else $parent[m] := -t$;
11.   define problem PL_m associated with node m
                  (constraints of PL_t plus $x_h \leq \lfloor val \rfloor$ if $child = 1$,
                  or $x_h \geq \lceil val \rceil$ if $child = 2$)
12.   solve problem PL_m, and let $\mathbf{x}^*$ be the optimal solution found ;
13. $LB[m] := \mathbf{c}^T\mathbf{x}^*$;
14.   **if** $(\mathbf{x}^*$ integer$)$ **and** $(\mathbf{c}^T\mathbf{x}^* < z_{OPT})$ **then**
                  begin /* update the optimal solution */
                      $\mathbf{x}_{OPT} := \mathbf{x}^*$; $z_{OPT} := \mathbf{c}^T\mathbf{x}^*$;
                      $Q := Q \setminus \{j \in Q \; : \; LB[j] \geq z_{OPT}\}$
                  end ;
15.   **if** $LB[m] < z_{OPT}$ **then**
                  begin
                      choose the fractional branching variable x_k^* ;
                      $vbranch[m] := k$; $value[m] := x_k^*$;
                      $Q := Q \cup \{m\}$
                  end
              end
          end
  end .

Figure 6.10: Branch-and-Bound Algorithm

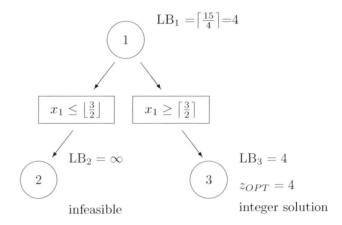

Figure 6.11: The branching tree of the numerical example

Example

Consider the following ILP problem (see Figure 6.11):

$$
\begin{cases}
\min & x_1 & +2x_2 & & \\
& x_1 & +x_2 & \leq & 4 \\
& x_1 & +4x_2 & \geq & 6 \\
& 8x_1 & -8x_2 & \geq & 3 \\
& x_1, & x_2 & \geq & 0 \text{ integer.}
\end{cases}
$$

Note that the formulation could be improved by replacing, in the third constraint, the right-hand side value 3 with 8.

Node 1 (root). Initially $z_{OPT} = +\infty$ (current optimal integer solution cost). By solving the continuous relaxation of the original problem, we obtain the optimal tableau:

		x_1	x_2	x_3	x_4	x_5
$-z$	$-\frac{15}{4}$	0	0	0	$\frac{3}{5}$	$\frac{1}{20}$
x_3	$\frac{11}{8}$	0	0	1	$\frac{2}{5}$	$\frac{3}{40}$
x_2	$\frac{9}{8}$	0	1	0	$-\frac{1}{5}$	$\frac{1}{40}$
x_1	$\frac{3}{2}$	1	0	0	$-\frac{1}{5}$	$-\frac{1}{10}$

The optimal solution has cost $z = \frac{15}{4}$, from which we obtain $\mathrm{LB}_1 = \lceil \frac{15}{4} \rceil = 4$. The optimal solution is $\mathbf{x}^* = \left[\frac{3}{2}, \frac{9}{8}, \frac{11}{8}, 0, 0 \right]^T$, for which we have three possible choices for branching variable x_h. Let us choose variable x_1, and generate nodes 2 and 3 in Figure 6.11.

Node 2. Starting from the optimal tableau of the parent node, we add constraint $x_1 \leq 1$, i.e., $x_1 + x_6 = 1$. In order to convert this constraint in canonical form, it is sufficient to obtain x_1 from row 3, and substitute. We obtain the augmented tableau:

		x_1	x_2	x_3	x_4	x_5	x_6
$-z$	$-\frac{15}{4}$	0	0	0	$\frac{3}{5}$	$\frac{1}{20}$	0
x_3	$\frac{11}{8}$	0	0	1	$\frac{2}{5}$	$\frac{3}{40}$	0
x_2	$\frac{9}{8}$	0	1	0	$-\frac{1}{5}$	$\frac{1}{40}$	0
x_1	$\frac{3}{2}$	1	0	0	$-\frac{1}{5}$	$-\frac{1}{10}$	0
x_6	$-\frac{1}{2}$	0	0	0	$\frac{1}{5}$	$\frac{1}{10}$	1
	1	1	0	0	0	0	1

By applying the dual simplex algorithm, we immediately discover that the current problem is infeasible ($\text{LB}_2 = +\infty$).

Node 3. Let us start again from the optimal tableau of the parent node, this time by adding constraint $x_1 \geq 2$, i.e., $x_1 - x_6 = 2$. Obtaining as before x_1 and substituting, we get:

		x_1	x_2	x_3	x_4	x_5	x_6
$-z$	$-\frac{15}{4}$	0	0	0	$\frac{3}{5}$	$\frac{1}{20}$	0
x_3	$\frac{11}{8}$	0	0	1	$\frac{2}{5}$	$\frac{3}{40}$	0
x_2	$\frac{9}{8}$	0	1	0	$-\frac{1}{5}$	$\frac{1}{40}$	0
x_1	$\frac{3}{2}$	1	0	0	$-\frac{1}{5}$	$-\frac{1}{10}$	0
x_6	$-\frac{1}{2}$	0	0	0	$-\frac{1}{5}$	$\left(-\frac{1}{10}\right)$	1
	2	1	0	0	0	0	-1

By solving the problem with the dual simplex, we obtain $\mathbf{x}^* = [2, 1, 1, 0, 5, 0]^T$ and the lower bound $\text{LB}_3 = \mathbf{c}^T \mathbf{x}^* = 4$. Since $\mathbf{x}^*$ is integer, we can update the value of the current optimal solution, obtaining $z_{OPT} = 4$. There is no need thus to generate the children of node 3 (this happens because $\mathbf{x}^*$ is integer—or, equivalently, because we have $LB_3 \geq z_{OPT}$ after the update of z_{OPT}). Since there are no other nodes to explore, the algorithm stops, and the current solution $\mathbf{x}_{OPT}$ is guaranteed to be optimal for the original ILP problem.

6.5   Branch-and-Cut

We will now consider a mixed *branch-and-bound/cutting-plane* technique, proposed in the 1990s by M. Padberg and G. Rinaldi to overcome some of the disadvantages of the two techniques.

The basic idea is simple: design a *branch-and-bound* algorithm in which, at each node t of the branching tree, some cuts are generated in the hope of obtaining an integer solution $\mathbf{x}_{PL_t}^*$, or even just a tighter (and hence more useful for fathoming) lower bound $\mathbf{c}^T\mathbf{x}_{PL_t}^*$. As soon as we see that these cuts become ineffective, cut generation stops and we switch to branching.

With respect to the pure *branch-and-bound*, there is the advantage of dynamically strengthening the formulation of the problem. With respect to the pure *cutting-plane* technique, instead, there is the possibility to hinder by means of branching the *tailing off* phenomenon (a long series of iterations without a substantial improvement of the current lower bound).

The practical realization of this idea is not immediate. For instance, suppose that we want to use Gomory's cuts within a *branch-and-bound* scheme. The cuts generated are obtained from the optimal tableau associated with the node of the current branching tree, hence their validity also depends on the branching constraints imposed. It follows that, in general, these cuts are only locally valid (they are only correct for the current node and for its possible child nodes).

Imagine instead of being able to generate *globally* valid cuts along the entire branching tree. It is then possible to memorize *all* cuts (as they are generated in the various nodes) in a global data structure, called the *cut pool*. When processing a new node of the branching tree, we start from formulation $\mathbf{Ax} = \mathbf{b}$, $\mathbf{x} \geq 0$, we add the branching constraints, and we solve the corresponding continuous relaxation obtaining solution $\mathbf{x}^*$ (say). If $\mathbf{x}^*$ is fractional, the pool is scanned for constraints violated by $\mathbf{x}^*$ to be added to the current formulation. We then solve the new formulation and we proceed in this way until $\mathbf{x}^*$ satisfies all the constraints of the pool. If $\mathbf{x}^*$ is still fractional, a convenient separation procedure can be called in order to identify new globally-valid cuts to be inserted into the pool.

The procedure described is called *branch-and-cut*. It is characterized by the presence of a global pool, and by one or more separation procedures which can generate globally valid cuts. Since the overall convergence is guaranteed by the branching mechanism, it is admitted that the separation procedures are of heuristic nature, i.e., it is allowed that, in some cases, they are not able to generate a cut.

The definition of effective separation procedures is one of the crucial points of the method. The ideal would be to have general-purpose procedures, i.e., procedures applicable to a generic ILP problem. The design of procedures of this kind is the subject of intense international research.

Another possibility is to study a *specific* class of ILP problems, exploiting their peculiar structure. Following this approach, the design of a *branch-and-cut* algorithm is divided into the following steps:

1. Identification of the structural properties of the ILP model under study.

2. Translation of the properties identified in terms of *classes of valid inequalities* (*polyhedral analysis*).

3. For each inequality class C, definition of effective procedures for the exact/heuristic solution of the following

   Separation Problem for the C class: *Given* $\mathbf{x}^*$, *identify (if there exists) an inequality* $\alpha^T\mathbf{x} \le \alpha_0$ *belonging to class C and such that* $\alpha^T\mathbf{x}^* > \alpha_0$.

In practice, for any class C we are interested in identifying several violated inequalities, chosen among those that maximize the degree of violation $\alpha^T\mathbf{x}^* - \alpha_0$. This typically accelerates the overall convergence of the *branch-and-cut* algorithm.

6.5.1   An Example: the Index Selection Problem

The main features of the *branch-and-cut* algorithm will be illustrated by means of a problem occurring in database design.

A relational database can be thought as a set of tables (data), plus *query* and update procedures. Typically, the data structure must be queried in real time, and therefore the response time to answer a query is a determining factor in the choice of the method with which to organize the information. The answer to a query involves scanning data, an operation that can be accelerated if the records are kept organized (using some keys) by means of one or more indices.

The response time to a given query is therefore a function of the index used. On the other hand, each index has a fixed cost (in terms of computing time) related to the periodic updating of data, and it occupies memory on its own.

As an example, consider the following numerical case. There are $m = 6$ queries and $n = 5$ potential indices. There is a dummy index, index 0, the use of which actually corresponds to the sequential scanning of data. The following table quantifies the response costs (computing times) to each query, using one of the indices provided; it is assumed that it is not possible to use more than one index for the same query.

query	index 0	index 1	index 2	index 3	index 4	index 5
1	6200	1300	6200	6200	6200	6200
2	2000	900	700	2000	2000	2000
3	800	800	800	800	800	800
4	6700	6700	6700	1700	6700	2700
5	5000	5000	5000	2200	1200	4200
6	2000	2000	2000	2000	2000	750

The fixed cost and the size in terabyte (TB) of the indices are the following:

	index 1	index 2	index 3	index 4	index 5
fixed cost	200	1200	400	2400	250
size	10	5	10	8	6

The total index storage space available is $D = 19$ TB.

A feasible solution corresponds to any subset of indices having size not greater than D (they are the indices that we actually want to create). The overall cost of the solution is obtained by summing the fixed costs and the costs related to the m queries, according to the following considerations.

Consider, for instance, solution $S = \{1,5\}$, which occupies only 16 TB, and is thus feasible. Solution S does not create indices 2, 3 and 4, which must then be removed from the previous tables. For each query $i \in \{1,\ldots,m\}$ we have to decide which index $j \in S \cup \{0\}$ to select. For query 1 it will be convenient to select index 1 (1300 being the minimum cost on the row). Index 1 will also be used for query 2 (cost 900), while index 5 will be used for queries 4 (cost 2700), 5 (cost 4200) and 6 (cost 750). Instead, index 0 is used for query 3 (cost 800). The overall cost to answer to the m queries is 10.650. We need to add to this cost the fixed costs (200 and 250) related to the creation/updating of indices 1 and 5. The total cost of solution $S = \{1,5\}$ is 10.650 + 450 = 11.100.

The *Index Selection Problem, ISP* consists in identifying a feasible solution with minimum cost. In the numerical example, it is easy to verify that the optimal solution is precisely $S = \{1,5\}$, with cost 11.100.

Suppose we want to design a *branch-and-cut* algorithm for this problem. The first step consists in defining a valid ILP model. To that end, let:

- m = total number of queries;

- n = total number of potential indices;

- D = available memory size for the indices selected;

- c_j = positive fixed cost of index $j \in \{1,\ldots,n\}$;

- d_j = positive size of index $j \in \{1,\ldots,n\}$;

- γ_{ij} = positive cost to answer query $i \in \{1,\ldots,m\}$ by means of index $j \in \{1,\ldots,n\}$.

The "natural" variables of the problem are, for all $j \in \{1,\ldots,n\}$:

$$y_j = \begin{cases} 1 & \text{if index } j \text{ is selected} \\ 0 & \text{otherwise.} \end{cases}$$

To express the fact that every query is associated with a selected index, we introduce the following auxiliary variables, defined for all $i \in \{1, \ldots, m\}$ and for all $j \in \{0, 1, \ldots, n\}$:

$$x_{ij} = \begin{cases} 1 & \text{if query } i \text{ uses index } j \\ 0 & \text{otherwise.} \end{cases}$$

Note that variables x_{ij} are defined also for $j = 0$ (dummy index).

A possible formulation is then the following.

$$\min \underbrace{\sum_{j=1}^{n} c_j y_j}_{\text{fixed cost}} + \underbrace{\sum_{i=1}^{m} \sum_{j=0}^{n} \gamma_{ij} x_{ij}}_{\text{query cost}} \tag{6.6a}$$

$$\underbrace{\sum_{j=1}^{n} d_j y_j \leq D}_{\text{storage usage}} \tag{6.6b}$$

$$\underbrace{\sum_{j=0}^{n} x_{ij} = 1}_{\text{only one index for each query}} \quad , \; i \in \{1, \ldots, m\} \tag{6.6c}$$

$$\underbrace{\sum_{i=1}^{m} x_{ij} \leq m y_j}_{y_j = 0 \;\Rightarrow\; x_{ij} = 0} \, , \; j \in \{1, \ldots, n\} \tag{6.6d}$$

$$0 \leq x_{ij} \leq 1 \text{ integer} \, , \; i \in \{1, \ldots, m\} \; ; j \in \{0, \ldots, n\} \tag{6.6e}$$

$$0 \leq y_j \leq 1 \text{ integer} \, , \; j \in \{0, \ldots, n\}. \tag{6.6f}$$

Constraint (6.6d) expresses the logic consistency of variables x_{ij} and y_j. In particular, when $y_j = 0$ the constraint imposes that $\sum_{i=1}^{m} x_{ij} = 0$ and hence $x_{ij} = 0$ for all $i \in \{1, \ldots, m\}$ (if index j is not selected, then it cannot be used by any query). Note that, due to the effect of factor m in the right-hand side, the constraint becomes inactive when $y_j = 1$ given that, in this case, index j (if convenient) can be used by all the m queries.

We now have an ILP formulation for the problem. If this formulation proves to be sufficiently tight, we can likely solve real-size problems (a few hundred indices and queries). Otherwise, it could be difficult to solve even very small instances, with only 15-20 indices and queries.

The ideal would be that solution $(\mathbf{x}^*, \mathbf{y}^*)$ of the continuous relaxation of the model had very few fractional components, and in any case a cost (6.6a) close to the optimal integer value (i.e., a tight lower bound). However, this hardly ever happens, mainly because of the *modeling mistake* we made when writing constraints (6.6d). Indeed, while condition "y_j integer" is maintained, these constraints act as requested. Relaxing the integrality condition we have, however, that the model, in the attempt of minimizing the costs, will assign to y_j^* the smallest value compatible with constraint (6.6d), i.e., it will systematically define $y_j^* = \frac{1}{m} \sum_{i=1}^{m} x_{ij}^*$. Therefore, even assuming that the values x_{ij}^* are all integer, we will have $y_j^* = 1$ only if $\sum_{i=1}^{m} x_{ij}^* = m$, that is, only in the (unlikely) case that a selected index j is then used by *all* m queries. In other words, coefficient m of variable y_j in constraint (6.6d) causes values y_j^* to be typically far from 1. On the other hand, the integer solutions *must* have a certain number of variables $y_j = 1$, and hence they will be identified by the *branch-and-bound* algorithm only in the deep nodes of the branching tree, after a considerable number of branching operations. Note that there exist 2^k nodes with depth k: if the integer solutions can be found on average at depth $k = 15$ or 20, this means having approximately $2^{15} - 2^{20}$ nodes in the branching tree!

A first strengthening of constraints (6.6d) may be easily obtained as follows. Consider, e.g., the column associated with index $j = 1$ in the query cost table, and note that some costs γ_{ij} are equal to the corresponding cost γ_{i0} on column 0 (this happens because index j is related to a key not related to query i). In this case, we can fix $x_{ij} = 0$, given that any optimal solution has no interest in using index j for query i, since it can use at the same cost index 0. It is thus valid to set to zero, i.e., to eliminate from the model, all variables x_{ij} such that $\gamma_{ij} \geq \gamma_{i0}$. Therefore, for every index $j \in \{1, \ldots, n\}$ only variables x_{ij} with $i \in I_j$ are active, where

$$I_j := \{i \in \{1, \ldots, m\} : \gamma_{ij} < \gamma_{i0}\}$$

is the set of queries for which it is convenient to use index j rather than index 0.

In the numerical example, the active variables (listed column-wise) are $x_{10}, \ldots, x_{60}$, x_{11}, $x_{21}, x_{22}, x_{43}, x_{53}, x_{54}, x_{45}, x_{55}$ and x_{67}, as well as variables $y_1, \ldots, y_5$.

In addition to reducing the number of variables of the problem, this reduction allow us to write constraints (6.6d) as

$$\sum_{i \in I_j} x_{ij} \leq |I_j| y_j \ , \quad j \in \{1, \ldots, n\}. \tag{6.7}$$

These constraints are stronger than constraints (6.6d), since the coefficient of y_j has been decreased from m to $|I_j|$.

Consider now the continuous relaxation of the new model. In our numerical example, the optimal solution $(\mathbf{x}^*, \mathbf{y}^*)$ has the following non-zero components:

$$x_{20}^* = \tfrac{6}{10}, \ x_{30}^* = 1$$
$$y_1^* = \tfrac{7}{10}, \quad x_{11}^* = 1, \ x_{21}^* = \tfrac{4}{10}$$
$$y_3^* = 1, \quad x_{43}^* = x_{53}^* = 1$$
$$y_5^* = \tfrac{1}{3}, \quad x_{65}^* = 1$$

and a (rounded) cost equal to 8.940. As can be seen, there is a significant difference (gap = 2.160) between the lower bound 8.940 and the optimal integer value 11.100. By directly applying the *branch-and-bound* algorithm, a large number of subproblems are generated, and the necessary computational time becomes prohibitive.

Let us then study the specific properties of the ISP problem, trying to identify new classes of valid inequalities (*polyhedral analysis*). To this end, it is useful to carefully analyze the fractional solution of the example. A "fault" of this solution is that index 1 is selected at 70% ($y_1^* = \frac{7}{10}$), but is used at 100% for query 1 ($x_{11}^* = 1$). It seems natural, instead, to impose that a partially selected index is only partially used by the queries. This structural condition, specific to the problem under study, can be expressed by the linear constraint $x_{11} \leq y_1$. We have thus identified a first class C_1 of valid inequalities for our problem:

$$\textbf{Class } C_1: \quad x_{ij} \leq y_j \, , \, j \in \{1, \ldots, n\}, \, i \in I_j. \tag{6.8}$$

Constraints (6.8) are similar to constraints (6.7), since they impose the congruence relation "$y_j = 0 \Rightarrow x_{ij} = 0$". However, in the absence of the integrality constraint, the new constraints may considerably strengthen the initial formulation, as is the case in the numerical example considered.

Note that constraints (6.7) can be obtained from (6.8) by simply summing up the conditions $x_{ij} \leq y_j$ for all $i \in I_j$. Constraints (6.7) are then linearly implied (la Farkas) by constraints (6.8). We could think of substituting constraints (6.7) with (6.8) in the model. This, however, would create a model with a very large number of constraints, since each constraint (6.7) is replaced by $|I_j|$ different constraints. On the other hand, not all these new constraints are necessary for solving a *specific* numerical instance: in the example, constraint $x_{21} \leq y_1$, even if not imposed, is satisfied by the current solution. So it can be convenient to generate "on the fly" (i.e., during the execution of the algorithm) the constraints from class C_1, limiting ourselves to those violated by the current fractional solution. To that end, we have to solve the following:

Separation Problem for class C_1: *Given* $(\mathbf{x}^*, \mathbf{y}^*)$, *identify a pair (i,j) such that* $x_{ij}^* > y_j^*$ *(if any).*

Given that family C_1 contains only $\sum_{j=1}^{n} |I_j| \leq n \cdot m$ inequalities, this problem can be solved with little computational effort, by enumeration.

Separation Procedure for class C_1;
   begin
        for $j := 1$ **to** n **do**
            for each $i \in I_j$ **do**
                if $x_{ij}^* > y_j^*$ **then**
                    "constraint $x_{ij} \leq y_j$ is violated by $(\mathbf{x}^*, \mathbf{y}^*)$"
   end .

In the numerical example, the procedure identifies two violated constraints, namely $x_{11} \leq y_1$ and $x_{65} \leq y_5$. We add those two cuts to the current model, and re-optimize using the dual simplex algorithm obtaining an optimal solution with cost 9.900 (gap = 1.200) and with the following non-zero components:

$$x_{30}^* = 1, \ x_{60}^* = \tfrac{3}{4}$$

$$y_1^* = 1, \quad x_{11}^* = x_{21}^* = 1$$

$$y_3^* = \tfrac{3}{4}, \quad x_{43}^* = x_{53}^* = \tfrac{3}{4}$$

$$y_5^* = \tfrac{1}{4}, \quad x_{45}^* = x_{55}^* = x_{65}^* = \tfrac{1}{4}.$$

By applying the separation algorithm for class C_1, no violated cuts are generated, meaning that the current solution satisfies *all* constraints (6.8), even if we explicitly imposed only two of them. At this stage, we can perform the branching operation, or we can study the fractional point in order to search for a new class of valid inequalities.

Note that indices 1 and 3 cannot be chosen at the same time, as they exceed the available memory D. We then have condition $y_1 + y_3 < 2$, which can be strengthen into $y_1 + y_3 \leq 1$ given that y_1 and y_3 must be integer. This latter inequality is violated by the current fractional point, given that $y_1^* = 1$ and $y_3^* = \tfrac{3}{4}$. More generally, we have the following new class of valid constraints:

$$\textbf{Class } C_2 : \ \sum_{j \in S} y_j \leq |S| - 1 \ , \ \forall\, S \subseteq \{1, \dots, n\} \, , \ S \neq \emptyset \ : \ \sum_{j \in S} d_j > D. \tag{6.9}$$

Since there exist $2^n - 1$ non-empty subsets $S \subseteq \{1, \dots, n\}$, class (6.9) can contain a huge number of inequalities. This is certainly a positive feature: the more inequalities, the tighter the resulting formulation (as there is a better chance of identifying a violated cut). Obviously, the separation algorithm for class C_2 cannot just enumerate all subsets S, verifying that $\sum_{j \in S} d_j > D$ and $\sum_{i \in S} y_j^* > |S| - 1$: the computational time requested would be huge.

A more sophisticated approach consists in formulating the separation problem as an ILP problem of its own, that can be solved, for instance, by means of the *branch-and-bound* algorithm. To that end, for all $j \in \{1, \dots, n\}$ we can introduce the auxiliary variables

$$z_j \ = \ \begin{cases} 1 & \text{if } j \text{ belongs to } S \\ 0 & \text{otherwise,} \end{cases}$$

and reformulate the separation problem as the problem of identifying, if there exists, a vector $\mathbf{z} \in \{0,1\}^n$ such that:

$$w := \underbrace{\sum_{j=1}^{n} z_j}_{|S|} - \underbrace{\sum_{j=1}^{n} y_j^* z_j}_{\sum_{j \in S} y_j^*} < 1 \quad (\, 1 - w \text{ being the constraint violation}) \tag{6.10a}$$

$$\underbrace{\sum_{j=1}^{n} d_j z_j}_{\sum_{j \in S} d_j} \geq D + \varepsilon, \qquad \text{(validity of the constraint)} \tag{6.10b}$$

where $\varepsilon > 0$ is a sufficiently small value (for example, $\varepsilon = 1$ if values $d_1, \ldots, d_n$ and D are all integer). If we want to identify the most violated constraint (6.9), we can interpret (6.10b) as an objective function, and write the following ILP model for the separation problem:

$$w^* := \min \sum_{j=1}^{n} (1 - y_j^*) z_j$$

$$\sum_{j=1}^{n} d_j z_j \geq D + \varepsilon$$

$$0 \leq z_j \leq 1 \text{ integer}, \; j \in \{1, \ldots, n\}.$$

For this problem, known in the literature as the minimization form of the *knapsack problem*, particularly effective solution codes exist. Note that values y_j^* are *known* values between 0 and 1, hence coefficients $1 - y_j^*$ of the objective function are all nonnegative. In addition, simple considerations show that it is possible to reduce the number of variables z_j of the problem by setting $z_j = 1$ if $y_j^* = 1$ (variable z_j has null cost in the objective function), and $z_j = 0$ if $y_j^* = 0$ (every solution with $z_j = 1$ would have a value of the objective function not lower than $1 - y_j^* = 1$, hence it would correspond to a non-violated constraint (6.9)).

By solving this easy (with respect to the original problem) ILP model, we can identify subsets S associated with maximally violated constraints (6.9): if $w^* < 1$, this constraint is indeed violated, and can be added to the current continuous relaxation. Otherwise the current solution $(\mathbf{x}^*, \mathbf{y}^*)$ satisfies *all* potential $2^n - 1$ constraints (6.9).

Another possibility is to heuristically limit the choice of subsets S imposing $|S| \leq k$ for any value k set (for instance, $k = 3$ or 4), according to the following scheme:

Heuristic separation procedure for class C_2;
   begin
        let $J^* := \{j \in \{1, \ldots, n\} : y_j^* > 0\}$;
        for each $S \subseteq J^* : 1 \leq |S| \leq k$ **do**
           if $\sum_{j \in S} d_j > D$ **then** /* valid constraint */
               if $\sum_{i \in S} y_j^* > |S| - 1$ **then**
                  "constraint $\sum_{j \in S} y_j \leq |S| - 1$ is violated by $(\mathbf{x}^*, \mathbf{y}^*)$"
   end .

Note that the algorithm only considers subsets S made by indices j with $y_j^* > 0$ given that, as already observed, constraint (6.9) cannot be violated when $j \in S$ and $y_j^* = 0$.

In the numerical example, by applying the separation algorithm for $k = 2$, we obtain the violated constraint $y_1 + y_3 \leq 1$. By adding this constraint to the current formulation and re-optimizing, we obtain an optima lfractional solution with cost 10.880 (gap = 220), with the following non-zero components:

$$x_{30}^* = 1$$
$$y_1^* = 1, \quad x_{11}^* = x_{21}^* = 1$$
$$y_4^* = \frac{3}{8}, \quad x_{54}^* = \frac{3}{8}$$
$$y_5^* = 1, \quad x_{45}^* = x_{65}^* = 1, x_{55}^* = \frac{5}{8}.$$

The separation procedure for class C_1 does not produce violated cuts, as that for class C_2 with $k = 2$. Trying with $k = 3$ we identify, instead, subset $S = \{1, 4, 5\}$, which corresponds to the violated constraint $y_1 + y_4 + y_5 \leq 2$. By adding this constraint and re-optimizing we obtain a solution with cost 11.100 (gap = 0) and non-zero components

$$x_{30}^* = 1$$
$$y_1^* = 1, \quad x_{11}^* = x_{21}^* = 1$$
$$y_5^* = 1, \quad x_{45}^* = x_{55}^* = x_{65}^* = 1.$$

Since no fractional component exists, this solution is optimal and the *branch-and-cut* algorithm stops (in this simple example) without the need to perform any branching operation.

Obviously, it is necessary to verify that the procedure described is effective for bigger problems, and on real data. As a curiosity, we report a computational comparison between the *branch-and-bound* technique (without constraints of the classes C_1 and C_2) and the *branch-and-cut* technique. The results are average values on 10 real life problems; the times are expressed in CPU seconds and refers to a (very old!) computer SUN Sparc-2.

		Branch-and-Bound		Branch-and-Cut	
m	n	node	time	nodes	time
250	150	> 10.000	> 2310.3	1	1.3
375	225	> 10.000	> 3030.2	17	31.2
500	300	> 10.000	> 4397.0	6	32.3

The *branch-and-bound* algorithm could not solve to optimality any of the 30 instances considered, having exceeded the maximum size of the queue of nodes (10.000 nodes). The same instances were easily solved, in few seconds, using the *branch-and-cut* technique. This proves the usefulness, for these problems, of considering the generation of cuts during the algorithm: the computational time needed to process each node increases, but the overall number of nodes is drastically reduced (from more than 10.000 nodes to less than 10, on average).

Chapter 7

Graph Theory

Some important problems in graph theory will now be studied. For each problem a mathematical model is presented, and possible solution algorithms are shown. In addition, some fundamental concepts of computational complexity theory are briefly introduced.

7.1 Undirected Graphs

An *undirected graph* is a pair $G = (V, E)$ in which V is a finite set and E is a family of unordered pairs of elements of V. The elements of V are called *vertices* (or *nodes*) of G, while the elements of E are called *edges* of G.

In the following, we will always denote by $n := |V|$ and $m := |E|$ the number of vertices and of edges of G, respectively.

In some cases, G is edge-*weighted* (or vertex-weighted), in which case a function $c : E \to \Re$ (or $w : V \to \Re$, respectively) is specified.

Typically $V = \{1, \ldots, n\}$, and the edges have the form $\{i, j\}$ with $i, j \in V$; for typographical reasons, we will use the equivalent notation $[i, j] \equiv [j, i]$. In Figure 7.1(a) a graph $G = (V, E)$ is shown with $V = \{1, 2, 3, 4, 5, 6\}$ and edges $[1, 4]$, $[1, 3]$, $[2, 3]$, $[2, 3]$, $[2, 4]$, $[2, 5]$, $[3, 4]$, $[4, 4]$. Note that edge $[2, 3]$ appears twice: in this case, we talk about *multiple* or *parallel* edges. Graphs without multiple edges are called *simple*. In some cases, edges of the type $[i, i]$ (called *loops*, as edge $[4, 4]$ in the example) are not allowed.

We say that edge $[i, j]$ *connects* (or is *incident to*) i and j. Two vertices are said to be *adjacent* if there exists an edge connecting them. Two edges are said to be *adjacent* (or *consecutive*) if they have a vertex in common.

The number of edges incident on a vertex v is called the *degree* of the vertex, and is usually denoted by $d_G(v)$. A vertex v is said to be *isolated* if $d_G(v) = 0$ (vertex 6 in Figure 7.1(a)), or *pendant* if $d_G(v) = 1$ (vertex 5 in Figure). It is easy to see that $\sum_{v \in V} d_G(v) = 2|E|$, hence it is always an even number.

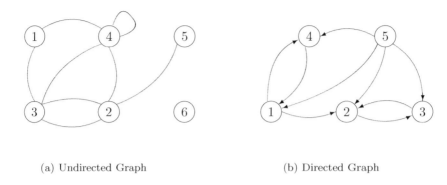

(a) Undirected Graph   (b) Directed Graph

Figure 7.1: Graphs

Graph G is said to be *complete* if it contains all possible edges, i.e., if $E = \{[i,j] : i,j \in V, i \leq j\}$.

A graph $G' = (V', E')$ is said to be a *subgraph* of $G = (V, E)$ if $V' \subseteq V$ and $E' \subseteq E$. If E' is the set of all edges of G with both endpoints in V', we say that G' is the subgraph of G *induced* by V'. A *partial graph* $G' = (V', E')$ of $G = (V, E)$ is a subgraph with $V' = V$: it has the same vertices of G but only a subset of its edges.

A *path* of G is a sequence of consecutive edges $e_1, e_2, \ldots, e_k \in E$ of the type: $e_1 = [v_1, v_2], e_2 = [v_2, v_3], \ldots, e_k = [v_k, v_{k+1}]$. We say that the path *connects* v_1 and v_{k+1}, *visiting* in sequence the *intermediate vertices* $v_2, v_3, \ldots, v_k$. The path is a *cycle* if $v_1 = v_{k+1}$. A path/cycle is said to be *elementary* if $e_i \neq e_j \ \forall \, i \neq j$, i.e., if it never uses the same edge twice. An elementary path/cycle is said to be *simple* if it never visits the same vertex twice, except for vertex $v_1 = v_{k+1}$ in the case of a cycle. In the example of Figure 7.2(a), the path shown is elementary but not simple, given that it visits vertex 2 twice.

Vertex v is said to be *connected* to another vertex w if there exists a path connecting them; note that the property is symmetric. For instance, in Figure 7.1(a) vertex $v = 1$ is connected to any other vertex apart from vertex $w = 6$. By definition, every vertex v is connected to itself, even if $[v, v] \notin E$.

A graph is said to be *connected* if vertices v and w are connected for all $v, w \in V$. The connectivity relationship induces a partition of the vertices of the graph, defined by its *connected components*, i.e., by the the maximal subsets of vertices that induce connected subgraphs. For instance, the connected components of the graph in Figure 7.1(a) are $\{1, 2, 3, 4, 5\}$ and $\{6\}$.

A *cut* of G is a set of edges of the type

$$\delta_G(S) := \{[i,j] \in E : |S \cap \{i,j\}| = 1\},$$

where S is the subset of vertices that induces the cut (i.e., the cut contains all edges with one endpoint in S and the other in $V \setminus S$). When unambiguous, we will write $\delta(S)$ instead of $\delta_G(S)$, and $\delta(v)$ instead of $\delta(\{v\})$.

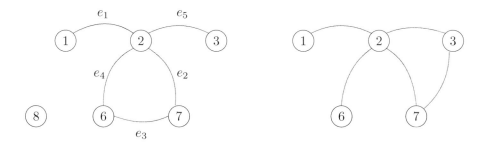

(a) Path between 1 and 3 through 2,7,6,2   (b) Connected Graph

Figure 7.2: Connectivity

In Figure 7.1(a) we have, for instance, cut $\delta(\{1,2,3\}) = \{[1,4],[2,4],[3,4],[2,5]\}$, while $\delta(6) = \emptyset$.

It is easy to verify that G is connected if and only if $\delta(S) \neq \emptyset$ for all $\emptyset \subset S \subset V$. More generally, *Menger's theorem* states that, given two vertices s and t, there exist k edge-disjoint paths connecting them if and only if $|\delta(S)| \geq k$ for all $S \subset V$ such that $s \in S$, $t \notin S$.

For all $S \subseteq V$, we use notation

$$E_G(S) := \{[i,j] \in E \; : \; i \in S \,, \, j \in S\}$$

to represent the set of edges of G with both endpoints in S. Again, we will write $E(S)$ instead of $E_G(S)$ unless it create ambiguous interpretations.

A partial graph $G' = (V, E')$ of G is called a *forest* if it is *acyclic*, i.e., it does not contain any cycle. A forest is *maximal* if every edge in $E \setminus E'$ forms a cycle with the edges in E'; in this case, it is easy to see that G' and G have the same connected components.

A maximal connected forest, if it exists, is called *spanning tree*, or simply *tree*. It is easy to verify that every tree has exactly $|V| - 1$ edges. In addition, there exists a tree in G if and only if G is connected. Otherwise, every maximal forest is formed by a collection of partial trees (one for each connected component).

A graph is said to be *bipartite* if there exists a partition (V_1, V_2) of V such that each edge $[i,j] \in E$ connects a vertex $i \in V_1$ to a vertex $j \in V_2$. It is easy to prove that G is bipartite if and only if it does not contain *odd cycles*, i.e., cycles with an odd number of edges; see Figure 7.3.

A graph is said to be *planar* if it can be drawn on the Euclidean plane in such a way that the lines representing the edges do not "cross" each other outside of the vertices. In Figure 7.4 two non-planar graphs are represented, called K_5 and $K_{3,3}$. A famous characterization by Kuratowski states that G is planar if and only if it does not contain a subgraph that derives from K_5 or $K_{3,3}$ by adding zero o more vertices along each edge.

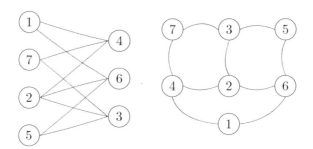

Figure 7.3: Bipartite graph

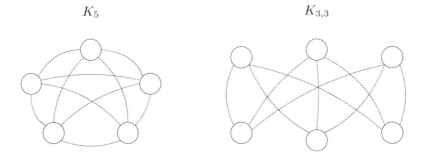

Figure 7.4: Non-planar graph

An elementary path or cycle is said to be *Eulerian* if it visits all edges of the graph exactly once. A connected graph is considered *Eulerian* if there exists an Eulerian cycle. A famous Euler's theorem says that G is Eulerian if and only if it is connected and each vertex $v \in V$ has even degree $d_G(v)$.

A simple path (or cycle) is said to be *Hamiltonian* if it visits all vertices of the graph exactly once. G is a *Hamiltonian graph* if it contains a Hamiltonian cycle. There are no known "simple" conditions to verify whether a given graph G is Hamiltonian. Obvious necessary (but not sufficient) conditions are that G must be connected, that $d_G(v) \geq 2$ $\forall\, v \in V$, and more generally that $|\delta(S)| \geq 2$ for all $\emptyset \subset S \subset V$.

A *clique* is a subgraph $G' = (V', E')$ of G in which every pair of vertices in V' is connected by an edge. By definition, $G_v := (\{v\}, \emptyset)$ is a clique for all $v \in V$. A clique G' is *maximal* if V' cannot be enlarged, i.e., no clique $G'' = (V'', E'')$ with $V'' \supset V'$ exists. A maximal clique $G' = (V', E')$ is *maximum* if there is no clique $G'' = (V'', E'')$ with $|V''| > |V'|$.

A *stable set* (or *independent set*) of G is a subgraph $G' = (V', E')$ induced by V' such that $E' = \emptyset$ (there exists no edge connecting two vertices in V'). As with cliques, it is possible to define the concepts of maximal and maximum stable sets.

7.2 Directed Graphs

A *directed graph* (or *digraph*) is a pair $G = (V, A)$ in which V is a finite set of *vertices* (or *nodes*) and A is a family of ordered pairs of vertices, called *arcs*. Hence, the arcs are pairs of the type (i, j) with $i, j \in V$. The order in which the two vertices appear is relevant, i.e., $(i, j) \neq (j, i)$. We say that arc $(i, j) \in A$ *leaves* vertex i and *enters* vertex j; vertices i and j are then called *tail* and *head* of the arc (i, j), respectively. In Figure 7.1(b) a directed graph is shown with 5 vertices; note that $(5, 4) \in A$ but $(4, 5) \notin A$.

As before, we will denote by n and m the number of vertices and arcs of graph G, respectively.

Many other definitions valid for undirected graphs also apply to directed graphs. We hence talk about simple directed graphs (i.e., without multiple arcs), loops, directed paths and directed cycles (otherwise called *circuits*), Hamiltonian and Eulerian paths/circuits, and so on.

In particular, a *directed path* is a sequence $a_1, a_2, \ldots, a_k$ of consecutive arcs of the type $a_1 = (v_1, v_2), a_2 = (v_2, v_3), \ldots, a_k = (v_k, v_{k+1})$. The path *starts* from v_1 and *reaches* v_{k+1}, passing through the *intermediate vertices* $v_2, \ldots, v_k$. The existence of such a path makes v_{k+1} *reachable* from v_1, and let v_1 *reach* v_{k+1}. The concept of reachability is hence the directed extension of the concept of connectivity. The property is obviously not symmetrical: in the example of Figure 7.1(b), vertex 4 cannot reach vertex 5.

For all $v \in V$ we will denote by Γ_v^+ the set of vertices *reachable* starting from v, and with Γ_v^- the set of vertices starting from which v can be reached. By definition, $v \in \Gamma_v^+ \cap \Gamma_v^-$ for all $v \in V$.

A directed graph G is said to be *strongly connected* if $\Gamma_v^+ = V$ for all $v \in V$, i.e., if there exists a path from v to w (and hence also from w to v) for all pairs $v, w \in V$. The relation induces a partition of V in its *strongly connected components* (maximal subsets of vertices that induce strongly connected subgraphs). A *directed cut* is a subset of arcs of the type:

$$\delta_G^+(S) := \{(i, j) \in A : i \in S, j \notin S\}$$

or of the type

$$\delta_G^-(S) := \{(i, j) \in A : i \notin S, j \in S\}$$

for some $S \subset V$; note that $\delta_G^+(S) \equiv \delta_G^-(V \setminus S)$. As usual we will omit, if possible, subscript G and we will write $\delta^+(v)$ and $\delta^-(v)$ instead of $\delta^+(\{v\})$ and $\delta^-(\{v\})$, respectively.

It is easy to prove that G is strongly connected if and only if $\delta^+(S) \neq \emptyset$ for all $\emptyset \subset S \subset V$. As in the undirected case, Menger's theorem states that there exist k directed arc-disjoint paths from an assigned vertex s to another assigned vertex t, if and only if $|\delta^+(S)| \geq k$ for all $S \subset V$ with $s \in S$ and $t \notin S$.

Given $S \subseteq V$, we will denote by

$$A_G(S) := \{(i, j) \in A : i \in S, j \in S\}$$

the set of arcs inside S, omitting subscript G when possible.

An *arborescence* is a partial graph $G' = (V, A')$ of G with $|A'| = |V| - 1$ arcs, in which a special vertex (the *root*) can reach every other vertex of the graph. Intuitively, an arborescence can be obtained from an undirected tree by choosing a root vertex and "propagating" from it the direction of the arcs.

A directed and strongly connected graph is said to be *Eulerian* if there exists a directed circuit that uses all arcs of the graph exactly once. It is easy to extend Euler's result to the directed case: G is Eulerian if and only if G is strongly connected and $|\delta^-(v)| = |\delta^+(v)|$ for all $v \in V$ (every vertex has the same number of entering and leaving arcs).

A directed graph is *Hamiltonian* if there exists a simple circuit passing exactly once through all the vertices of the graph. As for the undirected case, there are no known "simple" characterizations of Hamiltonian directed graphs.

The concept of planarity is typical of the undirected case, as are the cliques and the stable sets, hence it can be extended to the directed case simply by ignoring the orientation of the arcs.

7.3   Graph Representation

Both directed and undirected graphs can be described by appropriate matrices with $0 - 1$ elements, which play an important role in the mathematical formalization of some optimization problems defined on graphs.

Definition 7.3.1 *The node-edge incidence matrix D of an undirected graph $G = (V, E)$ is the $|V| \times |E|$ matrix with elements*

$$d_{ij} := \begin{cases} 1 & \text{if the } j\text{-th edge is incident to vertex } i \\ 0 & \text{otherwise.} \end{cases}$$

For instance, the undirected graph of Figure 7.5(a) corresponds to the following node-edge incidence matrix:

$$
\begin{array}{c}
\begin{array}{cccccccc}
 & a & b & c & d & e & f & g
\end{array} \\
\begin{array}{c}
1 \\ 2 \\ 3 \\ 4 \\ 5
\end{array}
\left[
\begin{array}{ccccccc}
1 & 1 & 0 & 0 & 0 & 0 & 0 \\
1 & 0 & 1 & 1 & 1 & 0 & 0 \\
0 & 1 & 1 & 0 & 0 & 1 & 0 \\
0 & 0 & 0 & 1 & 0 & 1 & 1 \\
0 & 0 & 0 & 0 & 1 & 0 & 1
\end{array}
\right]
\end{array}
$$

Note that every column of D has exactly two elements different from zero.

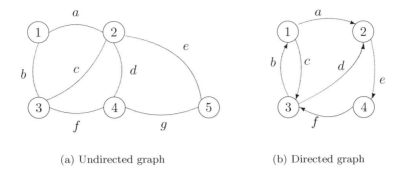

(a) Undirected graph (b) Directed graph

Figure 7.5: Graph representation

The node-edge incidence matrix D of a graph G is TUM if and only if G is bipartite. Indeed, if G is bipartite with vertex partition (V_1, V_2), then total unimodularity derives from Theorem 6.2.3 choosing $I_1 = V_1$ and $I_2 = V_2$. If G is not bipartite, then it contains at least one odd cycle. Let then C be the smallest odd cycle in G, and consider the submatrix Q of D associated with C:

$$
Q = \begin{bmatrix} 1 & 1 & 0 & 0 & 0 \\ 0 & 1 & 1 & 0 & 0 \\ 0 & 0 & 1 & 1 & 0 \\ 0 & 0 & 0 & 1 & 1 \\ 1 & 0 & 0 & 0 & 1 \end{bmatrix} \Rightarrow det(Q) = \pm 2 \Rightarrow D \text{ not TUM.}
$$

Definition 7.3.2 *The node-edge incidence matrix D of a directed graph $G = (V, A)$ is the $|V| \times |A|$ matrix with elements*

$$
d_{ij} := \begin{cases} 1 & \textit{if the j-th arc leaves vertex i} \\ -1 & \textit{if the j-th arc enters vertex i} \\ 0 & \textit{otherwise.} \end{cases}
$$

For instance, the node-edge incidence matrix associated with the directed graph of Figure 7.5(b) is:

$$
\begin{array}{c} \\ 1 \\ 2 \\ 3 \\ 4 \end{array}
\begin{array}{cccccc} a & b & c & d & e & f \\ \end{array}
\begin{bmatrix} 1 & -1 & 1 & 0 & 0 & 0 \\ -1 & 0 & 0 & -1 & 1 & 0 \\ 0 & 1 & -1 & 1 & 0 & -1 \\ 0 & 0 & 0 & 0 & -1 & 1 \end{bmatrix}
$$

Every column of the incidence matrix of a directed graph has exactly one $+1$ element and one -1 element in every column, hence it is TUM by Theorem 6.2.3 (choose partition $I_1 = V$, $I_2 = \emptyset$). In addition, at least one row of the matrix can be obtained as the linear combination of the other. One can indeed prove that $rank(D) = |V| - k$, where k is the number of connected components of the undirected graph obtained from G by eliminating the orientation of its arcs.

Another method to represent a graph is by means of a square $|V| \times |V|$ matrix, called *adjacency matrix* (of the nodes).

Definition 7.3.3 *The* adjacency matrix Q *of a simple graph* $G = (V, E)$ *is the symmetric* $|V| \times |V|$ *matrix with elements*

$$q_{ij} := \begin{cases} 1 & \text{if } [i, j] \in E \\ 0 & \text{otherwise.} \end{cases}$$

The adjacency matrix of the example in Figure 7.5(a) is:

$$
\begin{array}{c}
 \\ 1 \\ 2 \\ 3 \\ 4 \\ 5
\end{array}
\begin{array}{ccccc}
1 & 2 & 3 & 4 & 5 \\
\left[\begin{array}{ccccc}
0 & 1 & 1 & 0 & 0 \\
1 & 0 & 1 & 1 & 1 \\
1 & 1 & 0 & 1 & 0 \\
0 & 1 & 1 & 0 & 1 \\
0 & 1 & 0 & 1 & 0
\end{array}\right]
\end{array}
$$

Definition 7.3.4 *The* adjacency matrix Q *of a simple directed graph* $G = (V, A)$ *is the* $|V| \times |V|$ *matrix with elements*

$$q_{ij} := \begin{cases} 1 & \text{if } (i, j) \in A \\ 0 & \text{otherwise.} \end{cases}$$

The adjacency matrix of the directed graph of Figure 7.5(b) is:

$$
\begin{array}{c}
 \\ 1 \\ 2 \\ 3 \\ 4
\end{array}
\begin{array}{cccc}
1 & 2 & 3 & 4 \\
\left[\begin{array}{cccc}
0 & 1 & 1 & 0 \\
0 & 0 & 0 & 1 \\
1 & 1 & 0 & 0 \\
0 & 0 & 1 & 0
\end{array}\right]
\end{array}
$$

In the case of graphs with multiple edges/arcs, value q_{ij} is the number of times in which the given edge/arc appears in the graph. In the case of simple edge/arc-weighted graphs, it is instead possible to substitute value 0 with a conventional value, for instance $+\infty$, and value 1 with the value of the corresponding weight.

From a computer science point of view, a graph can always be represented by storing the scalar $n := |V|$ and an array of $m := |E|$ ($m := |A|$ for directed graphs) records, each of which stores the first and the second endpoints of an edge/arc, and possibly its weight. In some cases, a different data structure is however more convenient, which allows visiting the graph more efficiently.

A commonly-used data structure for undirected graphs stores, for all $v \in V$, its *star* $\delta(v)$. Note that every edge $[i, j]$ is then stored twice, given that $[i, j] \in \delta(i) \cap \delta(j)$. Usually the edges of a graph are stored in the array so that the edges in $\delta(1)$ occupy the first $|\delta(1)|$ positions, those of $\delta(2)$ the following $|\delta(2)|$ positions, and so on. Indicating with $FROM[h]$ and $TO[h]$ the two endpoints of the h-th edge, in the example of Figure 7.5(a) we would have:

	1	2	3	4	5	6	7	8	9	10	11	12	13	14
$FROM$	1	1	2	2	2	2	3	3	3	4	4	4	5	5
TO	2	3	1	3	4	5	1	2	4	2	3	5	2	4

$\delta(1) \quad \delta(2) \quad \delta(3) \quad \delta(4) \quad \delta(5)$

	1	2	3	4	5	6
$FIRST$	1	3	7	10	13	15

Note that the information $FROM[1, \ldots, m]$ is redundant, provided that we introduce the "pointer" vector $FIRST[1, \ldots, n+1]$ which identifies the position of the first edge in each star, with the convention $FIRST[1] = 1$, $FIRST[n + 1] = 2m + 1$, while $FIRST[v] = FIRST[v + 1]$ if $\delta(v) = \emptyset$. In this way, for all $v \in V$ the edges in $\delta(v)$ can be easily identified by just scanning positions $h = FIRST[v], FIRST[v]+1, \ldots, FIRST[v+1]-1$.

A similar data structure can be used for directed graphs, storing for each $v \in V$ the corresponding *forward star* $\delta^+(v)$ and/or the corresponding *backward star* $\delta^-(v)$.

Particular graphs can sometimes be stored more compactly. In particular, a simple directed path from s to t can be stored in an array $pred[1, \ldots, n]$ in which the elements are defined as follows:

$$pred[j] := \begin{cases} 0 & \text{if } j \text{ is not visited by the path} \\ s & \text{if } j = s \\ i & \text{if arc } (i, j) \text{ belongs to the path.} \end{cases}$$

In a similar way, it is possible to store an arborescence of root s, representing it as a set of paths still by means of a single vector $pred[1, \ldots, n]$ in which $pred[j] = i$ if arc (i, j) belongs to the arborescence, while $pred[s] = s$ for the root.

Obviously, every undirected tree can be stored as arborescence, by randomly choosing the root and orientating the edges in an appropriate way.

7.4   Computational Complexity Theory

The computational burden (in terms of computational time and storage usage) required to solve a problem depends on the algorithm used. Computational complexity theory tries to answer questions like: how efficient is a given algorithm? How *inherently* difficult is a given problem?

An *instance* of a problem is a specific case of the problem itself. For example, an instance of the problem "sort an array of n positive integer elements $c_1, c_2, \ldots, c_n$" can be associated with the data $n = 5, c_1 = 2, c_2 = 1, c_3 = 7, c_4 = 2, c_5 = 10$.

The *size* of an instance is the number of bits needed to encode it. In the case of the problem of sorting n integer values c_j, this size is equal to $\lceil \log_2 n \rceil$ to store value n, plus $\sum_{j=1}^{n} \lceil \log_2 c_j \rceil$ to store the n values $c_j > 0$.

In general, it is not necessary to precisely identify the size of an instance, but rather to know its order of magnitude. In this case, we typically use the following "Big O" notation:

Definition 7.4.1 *Given two real functions f and g defined on the same domain, we have that $f(x) = O(g(x))$ if there exist two constants k_1 and k_2 such that $f(x) \leq k_1 g(x) + k_2$ for all x in the domain.*

If, as often happens, f and g are monotonically increasing functions of the domain $\Re_+$, then $f(x) = O(g(x))$ if and only if $\lim_{x \to +\infty} \frac{f(x)}{g(x)}$ is a constant finite value. For instance, functions $f(x) = x^2 + 3x + 1$, $f(x) = 2x^2 + 5$, $f(x) = x + \log x$ are all $O(x^2)$ functions.

A function $f(x) \leq constant$ for all x is called $O(1)$.

In the case of the sorting problem, therefore, the size of an instance is $O(n \log L)$, where $L := \max\{c_1, c_2, \ldots, c_n\}$. If, on the other hand, the sorting problem were defined as follows: "Sort n integer values $c_1, \ldots, c_n$, with $|c_j| < 2^{31}$ for all $j = 1, \ldots, n$", the size of an instance would simply be $O(n)$, given that by hypothesis each value c_j can be stored with a fixed number of bits (32 bits). Often this latter hypothesis is implicitly made in the formulation of the problem; in these cases, the size of an instance coincides with the number of its data. Careful, though: by appropriately "packing" the data, it is always possible to save an instance in a *single* binary string, but this does not allow us to state that the size of each instance is $O(1)$!

Given an algorithm to solve a problem, we aim at determining the computational complexity as a function of the size of the instance that is to be solved. Typically the study refers to complexity in terms of *computational time*, expressed as the number of *elementary operations* (comparisons, sums, subtractions, etc. between operands of fixed dimension) necessary to solve an assigned instance of the problem. Obviously, this definition has to refer to a particular computational problem and to a limited set of operations that can each be performed in constant time, for instance by a *Turing machine*. For the sake of simplicity, we will skip this important aspect, assuming that our computational device is any deterministic and sequential computer.

The computational time also depends on the instance considered: a sorting algorithm like bubble-sort needs $O(n)$ time in the best case, but $O(n^2)$ time in the worst case. Usually, complexity analysis refers to the *worst case*. Sometimes it is possible to carry out *average-case* analyses that consider the average computing time related to a sample of instances characterized by a certain data probability distribution, or *experimental* analyses on significant test problems.

Definition 7.4.2 *An algorithm is said to be* polynomial *if it needs, in the worst case, a computing time $f(d) = O(d^k)$, where d is the size of the instance and k is a fixed constant.*

Algorithms $O(n^2)$ and $O(n \log n)$ for sorting a vector of (small) integers are hence polynomial, as are the algorithms for inverting an $m \times m$ matrix of (small) integers, which require $O(m^3)$ time. On the other hand, the simplex algorithm for linear programming is not polynomial as it can be shown that, in the worst case, it requires about $\frac{n!}{m!(n-m)!}$ iterations. Even the *branch-and-bound* technique for integer linear programming is not polynomial, given that, still in the worst case, approximately 2^n nodes of the branching tree can be generated. The same applies to the *cutting-plane* and *branch-and-cut* techniques, which require, in the worst case, a computational time that increases exponentially with the size of the instance.

A polynomial algorithm is considered *efficient*—even if an algorithm $O(d^{17})$ is "efficient" in theory, but very little in practice!

It is possible to extend the complexity concept from the algorithm to the problem itself:

Definition 7.4.3 *A problem is said to be* polynomial *(or easy) if there exists a polynomial algorithm solving it.*

The problems of sorting a vector and inverting a matrix are obviously easy. More delicate is the matter dealing with "probably hard" problems, for which no polynomial algorithm is known yet, but nothing excludes that such an algorithm exists. An emblematic example is linear programming: considered a hard problem until 1979, it became solvable in polynomial time by means of an innovative algorithm (the *ellipsoid method*) by the Russian researcher L.J. Khachiyan. It is clear then that the distinction between easy and hard problems is related to the current state of the art, although we can probably assume that some problems are inherently harder than others.

Computational complexity theory seeks to solve, at least partially, the above mentioned ambiguities by defining an inherent complexity hierarchy of the problems. Some of the hallmarks of this theory are the concepts of classes $P, NP, \text{co-}NP$ and of NP-complete problem, which will be sketched in what follows.

Complexity theory usually refers to *decision* problems (rather than optimization problems), in which the algorithm has to provide as output a yes/no answer to a well formulated question. Here are some simple examples:

(1) Given an integer matrix A and an integer vector $\mathbf{b}$, does there exist a solution $\mathbf{x} \geq 0$ of the system $A\mathbf{x} \leq \mathbf{b}$?

(2) Given an integer matrix A, two integer vectors $\mathbf{c}$ and $\mathbf{b}$, and an integer scalar L, does there exist $\mathbf{x} \geq 0$ such that $A\mathbf{x} \leq \mathbf{b}$, $\mathbf{c}^T\mathbf{x} \leq L$?

(3) Is a given undirected graph $G = (V, E)$ connected (or bipartite, or Eulerian, or planar) ?

(4) Given an undirected graph $G = (V, E)$ and a value $L \in \{1, \ldots, |V|\}$, does there exist a clique $G' = (V', E')$ of G with $|V'| \geq L$?

(5) Is a given undirected graph $G = (V, E)$ Hamiltonian?

(6) Is a given undirected graph $G = (V, E)$ *non*-Hamiltonian?

Note that problem (4) is tightly connected to the following optimization problem:

(4') Given $G = (V, E)$, identify the cardinality $z^* := |V'|$ of a maximum clique $G' = (V', E')$ of G.

Clearly, an algorithm to solve (4') can be used for problem (4) by simply checking that $z^* \geq L$. However, even the opposite is true: for all fixed L we can use, as a black-box, any algorithm for problem (4) to test if $z^* \geq L$ or $z^* < L$. By means of a binary search on value L, it is hence possible to identify value z^* by solving $O(\log |V|)$ times problem (4). It follows that (4) and (4') are in a way equivalent formulations of the same problem.

A similar approach can be applied to most optimization problems, which justifies the claim that decision problems may conveniently represent optimization problems as well.

For the first three problems considered, polynomial algorithms are known, i.e., problems (1), (2) and (3) are *certainly* polynomial. The first 5 problems share an important property: it is easy to certify the "yes" answer by exhibiting a *given* solution (found no matter how). For instance, for problem (5) a "yes" answer can be certified simply exhibiting a Hamiltonian cycle of G.

Definition 7.4.4 *A polynomial certificate is defined as an auxiliary piece of information that can be used to verify, in time polynomial in the size of the instance, the correctness of the answer for a given instance.*

Note that it is not important *how* this certificate is obtained, but only that it exists.

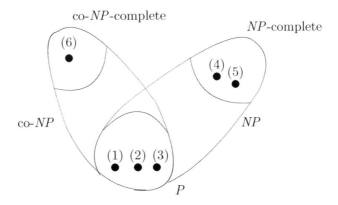

Figure 7.6: Classes of problems

Definition 7.4.5 *A decision problem belongs to the NP* (Nondeterministic Polynomial) class *if every of its instances with "yes" answer admits a polynomial certificate.*

It is easy to verify that problems (1)-(5) belong to the *NP* class, while problem (6) probably does not belong to it since one of its instances with "yes" would need a polynomial certificate to prove that the considered graph is *not* Hamiltonian, a certificate that up until now is unknown (the exhibition of all possible subsets of $|V|$ edges would be a *non*-polynomial certificate). Also the specular concept is possible:

Definition 7.4.6 *A decision problem belongs to the* co-*NP* (*NP* complete) class *if every of its instances with "no" answer admits a polynomial certificate.*

Hence problem (6) certainly belongs to the co-*NP* class, while probably problems (4) and (5) do not belong to this class.

Note that polynomial decision problems certainly belong to the $NP \cap$ co-*NP* class, given that every instance with "yes" or "no" answer can be certified simply by solving the problem in polynomial time, without the need of any auxiliary information (the certificate is empty).

We hence have the hierarchy shown in Figure 7.6, in which P is the class of polynomial problems. Note that, in theory, nothing excludes that $NP =$ co-*NP* or that $P = NP$, even though both cases are now considered extremely unlikely.

It is clear that there exist relation of the type "problem P_2 is at least as hard as problem P_1" among the various problems of the *NP* class. For instance, problem (2) is at least as hard problem (1), given that by choosing $\mathbf{c} = 0$ and $L = 0$ in (2) we obtain exactly problem (1). On the other hand, expressing condition $\mathbf{c}^T\mathbf{x} \leq L$ as an additional constraint in $\mathbf{A}\mathbf{x} \leq \mathbf{b}$, we can easily link (2) to problem (1). In other words, an algorithm for (1) can solve (2), and vice versa: the two problems are hence equivalent.

In general, the following concept of *polynomial reduction* among problems applies.

Definition 7.4.7 *Problem P_1 can be* reduced in polynomial time *to problem P_2 ($P_1 \propto P_2$) if there exists an algorithm to solve P_1 that:*

1. *uses as black-box procedure any algorithm for P_2;*

2. *is polynomial in the hypothesis that the algorithm for P_2 needs $O(1)$ time.*

In other words, an algorithm for P_1 exists that uses, as a black-box procedure, an algorithm for P_2 for a polynomial number of times. Note that this property is transitive: if $P_1 \propto P_2$ and $P_2 \propto P_3$, then $P_1 \propto P_3$. As an exercise, the reader can prove that $(1) \propto (2)$ and that $(2) \propto (1)$.

The easiest case is when the algorithm for P_1 uses the procedure for P_2 only once, according to the following scheme:

1. gets the input data of the instance of problem P_1 that we want to solve;

2. construct, in polynomial time, a suitable instance of P_2 that has the same (yes/no) answer of the instance of P_1;

3. solve the instance of P_2 thus obtained, also obtaining the answer for P_1.

When the above construction is possible, we say that P_1 *transforms* in polynomial time to P_2.

From the definition, it immediately follows that P_2 is at least as hard as P_1 whenever P_1 can be polynomially reduced to P_2, i.e.:

$$P_1 \text{ “hard” }, P_1 \propto P_2 \quad \Rightarrow \quad P_2 \text{ “hard”}.$$

Indeed, if there exists a polynomial algorithm for P_2, then this could be used as black-box procedure for solving P_1, still in polynomial time. It is then clear that the polynomial reduction relation induces a complexity hierarchy among problems of the *NP* class, which enables us to qualify some problems as the hardest of the entire class.

Definition 7.4.8 *A problem $P_2 \in NP$ is said to be NP-complete if $P_1 \propto P_2$ for all $P_1 \in NP$.*

It is possible to define in a similar way co-*NP* complete problems (see Figure 7.6).

The hallmark of *NP*-complete problems is that if anyone ever manages to solve *any of them* in polynomial time, *all* the other problems of the *NP* class would be solved in polynomial time, basically using the same algorithm, hence we would have $P = NP$. Since now it is believed that $P \neq NP$, it follows that it is very unlikely that an *NP*-complete problem can be solved in polynomial time.

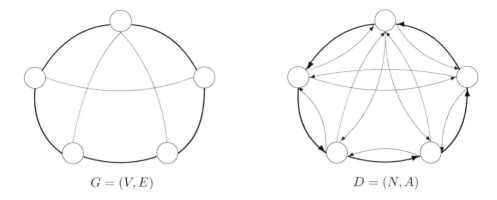

$$G = (V, E) \qquad\qquad D = (N, A)$$

Figure 7.7: Hamiltonian cycles and circuits

A fundamental result is that there indeed exist *NP*-complete problems, which was proved by S.A. Cook in 1971 for the *Boolean satisfiability* problem. Using the reduction mechanism starting from this problem it was subsequently possible to prove that hundreds and hundreds of problems, including problems (4) and (5), are *NP*-complete.

A technique to prove that a given decision problem P_* is *NP*-complete is the following:

1. Prove that $P_* \in NP$, establishing a polynomial certificate in the case of instance with "yes" answer;

2. Identify a problem P_2 "close" to P_* that is already known to be *NP*-complete, and prove that P_2 can be reduced in polynomial time to P_*. It follows that every other problem $P_1 \in NP$ can be polynomially reduced to P_* passing through P_2, given that $P_1 \propto P_2$ (since P_2 is *NP*-complete) and $P_2 \propto P_*$ (as proven).

As an example, consider the following problem

(7)   Given a directed graph $D = (N, A)$, does there exist a Hamiltonian circuit $D' = (N, A')$ in D?

Theorem 7.4.1 *Problem (7) is NP-complete.*

Proof: Problem $P_* \equiv (7)$ belongs to the *NP* class, given that one of its instances with "yes" answer can be certified by just exhibiting a Hamiltonian circuit in D.

Consider now problem $P_2 \equiv (5)$, asking for the existence of a Hamiltonian cycle in an undirected graph $G = (V, E)$. Given any instance of P_2, we can create in polynomial time a directed graph $D = (N, A)$ obtained from G by bi-orienting every of its edges, i.e., by defining $N := V$ and $A := \{(i,j), (j,i) : [i,j] \in E\}$; see Figure 7.7. It is easy then to verify that there exists a correspondence between the Hamiltonian circuits of D and the Hamiltonian cycles of G, i.e., G is Hamiltonian if and only if D is Hamiltonian. It follows that $P_2 \propto P_*$, given that an algorithm to solve problem P_2 on a given undirected graph $G = (V, E)$ can: (1) create in polynomial time the directed graph $D = (N, A)$; (2) solve problem P_* on D, by finding out if D (and hence G) is Hamiltonian.

Hence $P_2 \propto P_*$, and since P_2 is known to be NP-complete we have that P_* is NP-complete as well.   □

Finally, there exists another class of problems, the one of NP-*hard* problems. These are problems P_* not necessarily in NP, which have the characteristic that $P_1 \propto P_*$ for all $P_1 \in NP$ (hence the existence of a polynomial algorithm for P_* would imply that $P = NP$). The "optimization versions" of NP-complete problems belong to the class of NP-hard problems, as for instance the following *Traveling Salesman Problem*:

(8)   Given an undirected complete graph $G = (V, E)$ with integer weights $c_e \geq 0$ for all $e \in E$, identify a Hamiltonian cycle $G' = (V, E')$ in G with minimum cost $\sum_{e \in E'} c_e$.

7.5   Minimum Spanning Trees

Consider a given undirected graph $G = (V, E)$ with $n := |V|$ vertices and $m := |E|$ edges, and let $c : E \to \Re$ be a given *cost* (or *weight* or *length*) function, defined on the edges of G. For all $e \in E$ we will denote by c_e or by c_{ij} the cost of edge $e = [i, j]$, while for all $E' \subseteq E$ we will denote by $c(E') := \sum_{e \in E'} c_e$ the overall cost of the edges in E'.

The *Minimum Spanning Tree Problem* (MST) consists in identifying a tree $G_T = (V, T)$ in G with minimum cost $c(T)$. Obviously, the problem admits a solution if and only if G is connected.

An ILP model for the minimum spanning tree problem is the following. For all $e \in E$, consider the following decision variables:

$$x_e = \begin{cases} 1 & \text{if } e \text{ is chosen in the minimum spanning tree} \\ 0 & \text{otherwise.} \end{cases}$$

We then have the model:

$$\min \underbrace{\sum_{e \in E} c_e x_e}_{\text{total cost}} \tag{7.1a}$$

$$\underbrace{\sum_{e \in E} x_e = n - 1}_{n-1 \text{ chosen edges}} \tag{7.1b}$$

$$\underbrace{\sum_{e \in E(S)} x_e \leq |S| - 1}_{\text{cycle elimination condition}} \quad \forall S \subseteq V : S \neq \emptyset \tag{7.1c}$$

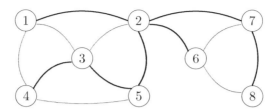

Figure 7.8: Graph and spanning tree

$$x_e \geq 0 \text{ integer} , \ e \in E. \tag{7.1d}$$

The $2^n - 1$ constraints (7.1c) guarantee the absence of cycles in the partial graph $G_T = (V, T := \{e \in E : x_e = 1\})$ associated with a solution $\mathbf{x}$ of the model. Note that constraint (7.1c) imposes $x_{ij} \leq 1$ in case $S = \{i, j\}$.

One can prove that the described formulation is ideal, i.e., that the continuous relaxation of model (7.1a)-(7.1d) does not have fractional vertices, even if the constraint matrix is not TUM. In this example we can see how the ideal formulation of a simple (and, as we will see, easy to solve) problem can involve an exponential number of constraints, showing that the complexity of a problem is not directly connected to the *number* of linear constraints necessary to described it by means of a linear programming model.

7.5.1 Efficient Algorithms

In the following, we will suppose that

(1) G is connected,

(2) $c_e \neq c_f \ \ \forall e, f \in E , \ e \neq f$.

Hypothesis (1) is needed to guarantee the existence of the minimum spanning tree: if G were not connected, the described algorithm would allow the identification of a *maximal forest* with minimum cost. Hypothesis (2) is needed to guarantee the *uniqueness* of the minimum spanning tree, which allow us to formulate more precisely its fundamental properties. In case condition (2) does not apply, the proposed algorithms would however still be correct, and would allow to identify *one* of the optimal solution of the problem.

Given a tree $G_T = (V, T)$ and an edge $e \notin T$, we will indicate with $C[T, e]$ the set of edges of the unique cycle contained in $T \cup \{e\}$.

Theorem 7.5.1 *Let $G_{T^*} = (V, T^*)$ be a minimum spanning tree. For any $e \in E$ we have that $e \in T^*$ if and only if there exists an $S \subset V$ such that $e = \arg\min\{c_f : f \in \delta(S)\}$.*

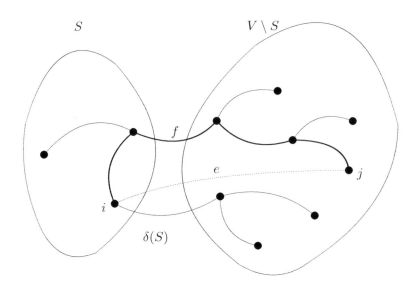

Figure 7.9: Cycle $C[T^*, e]$ in $T^* \cup \{e\}$

Proof: First, we will prove that the condition is sufficient, i.e., that "$e = \arg\min\{c_f : f \in \delta(S)\}$ for any cut $\delta(S) \Rightarrow e \in T^*$". Let us assume by contradiction that $e \notin T^*$. As shown in Figure 7.9, the set of edges $T^* \cup \{e\}$ contains a cycle $C[T^*, e]$ that "crosses" the cut $\delta(S)$ at least twice. Choose then any edge $f \in C[T^*, e] \cap \delta(S)$, $f \neq e$, and observe that $c_e < c_f$ given that $f \in \delta(S) \setminus \{e\}$. But then the set of edges $T := (T^* \setminus \{f\}) \cup \{e\}$ defines a tree $G_T = (V, T)$ with cost $c(T) = c(T^*) - c_f + c_e < c(T^*)$, which is impossible given that T^* is a minimum spanning tree.

Let us prove now that the condition is also necessary, i.e., that "$e \in T^* \Rightarrow$ there exists $S \subset V$ such that $e = \arg\min\{c_f : f \in \delta(S)\}$". To that end, it is sufficient to choose as set S one of the two connected components of the partial graph $G' = (V, T^* \setminus \{e\})$, and observe that if there existed $f \in \delta(S) \setminus \{e\}$ with $c_f < c_e$, then $G_{T^*} = (V, T^*)$ would not be a minimum spanning tree, given that the alternative tree $G_T = (V, T)$ with set of edges $T := (T^* \setminus \{e\}) \cup \{f\}$ would have cost $c(T) = c(T^*) - c_e + c_f < c(T^*)$.   □

In a similar way, it is also possible to prove that a given tree $G_{T^*} = (V, T^*)$ is a minimum spanning tree if and only if, for all $e \notin T^*$ and for all $f \in C[T^*, e] \setminus \{e\}$, we have $c_f \leq c_e$.

Theorem 7.5.1 suggests the scheme of Figure 7.10 to identify a minimum spanning tree $G_{T^*} = (V, T^*)$ in G.

Note that the set S defined at Step 3 must be the union of one or more connected components of the current partial graph $G_{T^*} = (V, T^*)$.

At each iteration, the algorithm adds to the current set T^* an edge e that we know to belong to the minimum spanning tree by Theorem 7.5.1. Obviously, this edge cannot

MST Algorithm;
   begin
1. $T^* := \emptyset$;
2.   **while** $|T^*| \neq n - 1$ **do**
        begin
3.   identify a set $S \subset V$ such that $\delta(S) \cap T^* = \emptyset$, and let
            $e := \arg\min\{c_f \; : \; f \in \delta(S)\}$;
4. $T^* := T^* \cup \{e\}$
        end
   end .

Figure 7.10: Prototype algorithm for minimum spanning tree

Prim-Dijkstra's Algorithm (1st version);
   begin
1. $T^* = \emptyset$;
     $S := \{1\}$;
2.   **while** $|T^*| \neq n - 1$ **do**
        begin
3.   identify edge $[v, h] \in \delta(S)$ of minimum cost, with $v \in S$ and $h \notin S$;
4. $T^* := T^* \cup \{[v, h]\}$;
        $S := S \cup \{h\}$
        end
   end .

Figure 7.11: Prim-Dijkstra's Algorithm

create cycles (given that, at Step 3, $\delta(S) \cap T^* = \emptyset$), and is well defined if there exists $S \subset V$ with $\delta(S) \cap T^* = \emptyset$, i.e., as long as $G_{T^*} = (V, T^*)$ is not connected.

The MST algorithm can be implemented in a number of ways, depending on how set S is chosen at Step 3.

7.5.2 Prim-Dijkstra's Algorithm

The most natural implementation of the MST algorithm is perhaps that of Prim-Dijkstra, in which at each iteration we choose the connected component S which contains an assigned vertex (for example vertex 1). The algorithm is shown in Figure 7.11.

As an example, consider the graph in Figure 7.12. By applying the procedure we obtain, iteration after iteration:

1. $S = \{1\}$, $v = 1$, $h = 4$

2. $S = \{1, 4\}$, $v = 1$, $h = 3$

3. $S = \{1, 4, 3\}$, $v = 4$, $h = 6$

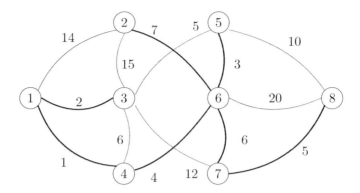

Figure 7.12: Example of Prim-Dijkstra's Algorithm

4. $S = \{1, 4, 3, 6\}$, $v = 6$, $h = 5$

5. $S = \{1, 4, 3, 6, 5\}$, $v = 6$, $h = 7$

6. $S = \{1, 4, 3, 6, 5, 7\}$, $v = 7$, $h = 8$

7. $S = \{1, 4, 3, 6, 5, 7, 8\}$, $v = 6$, $h = 2$

8. $S = \{1, 4, 3, 6, 5, 7, 8, 2\}$: STOP.

The computational complexity of the method depends on how, at Step 3, edge $[v, h]$ is identified. If we explicitly scan all the m edges of the graph, each iteration of the **while-do** loop needs $O(m)$ time, and the resulting algorithm has $O(m \cdot n)$ complexity, i.e., $O(n^3)$ in case of a complete graph with $m = \frac{n(n-1)}{2}$ edges. It is however possible to obtain a complexity of $O(n^2)$ if, at each iteration, we make appropriate use of the information already acquired in the previous iterations.

Consider the case of a complete graph $G = (V, E)$ with n vertices, stored by means of a symmetric $n \times n$ matrix whose element $c[i, j] = c[j, i]$ is the cost of edge $e = [i, j]$ (if G is not complete, define $c[i, j] := c[j, i] := +\infty$ for all $[i, j] \notin E$).

The algorithm described in Figure 7.13 uses, and iteratively updates, the following data structure:

- k = number of chosen edges;

- $flag[i] = \begin{cases} 1 & \text{if } i \in S \\ 0 & \text{otherwise} \end{cases}$ for all $i = 1, \ldots, n$;

- $L[j] = \min\{c_{ij} : i \in S\}$, defined for all $j \notin S$;

Prim-Dijkstra's Algorithm ($O(n^2)$ version);
  begin
1. $flag[1] := 1$;
     $pred[1] := 1$;
     for $j := 2$ **to** n **do** /* vertices $j \notin S$ */
       begin
         $flag[j] := 0$;
         $L[j] := c[1, j]$;
         $pred[j] := 1$
       end ;
2.   **for** $k := 1$ **to** $n - 1$ **do** /* select the $n - 1$ edges of the tree */
     begin
3. $min := +\infty$;
       for $j := 2$ **to** n **do** /* choose the minimum edge in $\delta(S)$ */
         if $(flag[j] = 0)$ **and** $(L[j] < min)$ **then**
           begin
             $min := L[j]$;
             $h := j$
           end ;
4. $flag[h] := 1$; /* extend S */
5.   **for** $j := 2$ **to** n **do** /* update $L[j]$ and $pred[j]$ for all $j \notin S$ */
         if $(flag[j] = 0)$ **and** $(c[h, j] < L[j])$ **then**
           begin
             $L[j] := c[h, j]$;
             $pred[j] := h$
           end
     end
  end .

Figure 7.13: Prim-Dijkstra's $O(n^2)$ Algorithm

- $pred[j] = \begin{cases} \arg\min\{c_{ij} \,:\, i \in S\} & \text{for all } j \notin S \\ \text{predecessor of } j \text{ in the minimum spanning tree} & \text{for all } j \in S \end{cases}$

At Step 5, set S includes the new vertex h, hence $L[j]$ and $pred[j]$ must be updated for all $j \notin S$. In output, the minimum spanning tree is defined on edges $[pred[j], j], j = 2, \ldots, n$.

The complexity of the algorithm may be evaluated as follows. The initialization at Step 1 needs $O(n)$ time. Steps 3 and 5 need $O(n)$ time each, and are performed $n - 1$ times. The complexity of the entire algorithm is thus $O(n^2)$, i.e., the best possible when the graph has $O(n^2)$ edges as in the case of complete graphs. The average efficiency of the algorithm can however be improved by explicitly storing, in a list, vertices $j \notin S$ (those with $flag[j] = 0$).

In the case of sparse graphs with $m \ll n^2/2$, it is convenient to use a different implementation of Prim-Dijkstra's algorithm, which requires $O(m \log n)$ time. The idea is to store vertices $j \notin S$ by means of a *priority queue* Q (for instance a *heap* queue), i.e., by means of a data structure that allow us to perform in $O(\log |Q|)$ time the following elementary operations:

- identification of element $h := \arg\min\{L[j] \,:\, j \in Q\}$;

- elimination from Q of an element j;

- insertion in Q of an element $j \notin Q$.

At Step 1 of Prim-Dijkstra's algorithm, queue Q can be initialized in $O(n \log n)$ time, by sequentially inserting elements $j = 2, \ldots, n$ (actually, this initialization can be performed in $O(n)$ time). At Step 3, vertex h can be identified and eliminated from Q in $O(\log n)$ time. As for Step 5, it is possible to proceed as follows. We scan edges $[h, j] \in \delta(h)$, stored in the star of vertex h as explained in the related section of page 131: if $(flag[j] = 0)$ *and* $(c[h, j] < L[j])$ we extract j from Q in $O(\log n)$ time, we update $L[j]$ and $pred[j]$ in $O(1)$ time, and then we insert j in Q with the new value of $L[j]$, still in $O(\log n)$ time. In this way, each edge $[u, v] \in E$ of the graph is considered only twice (when we scan $\delta(h)$ for $h = u$ and $h = v$), hence the overall number of updates is just $O(m)$.

The global complexity of the algorithm hence is $O(n \log n)$ for the initialization at Step 1, plus $O(m \log n)$ for the $O(m)$ updates at Step 5, i.e., $O(m \log n)$ in total. The details of this implementation are left as an exercise for the reader.

7.5.3   Kruskal's Algorithm

A different implementation of the MST prototype algorithm of Figure 7.10 can be obtained by *first* choosing the edge e with minimum cost, and *then* verifying the existence of a $S \subset V$ such that $e \in \delta(S)$. More precisely, edges $[i, j] \in E$ are iteratively scanned according to increasing costs c_{ij}. At each iteration, let $[i, j]$ be the current edge, and let C_i and C_j be the connected components of the current graph $G_{T^*} = (V, T^*)$ that contain

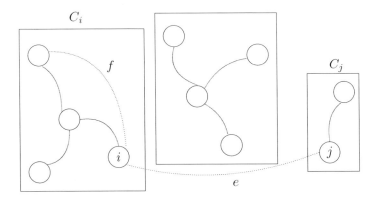

Figure 7.14: Connected components of $G_{T^*} = (V, T^*)$

i and j, respectively. If $C_i = C_j$ (as, for instance, for edge f in Figure 7.14), then the current edge cannot be selected as it forms a cycle with the edges already present in T^*. If, instead, $C_i \neq C_j$ (as, for instance, for edge e in Figure 7.14) then, by construction, edge $[i, j]$ is the minimum cost edge in $\delta(S)$, where $S = C_i$ or $S = C_j$. In this latter case, it is hence allowed to add $[i, j]$ to T^*, update the connected components of G_{T^*} by combining C_i and C_j in a single component, and moving to the next edge in the ordered sequence.

The resulting algorithm is hence of the *greedy* type, i.e., at each iteration it performs the choice that is considered the best (selecting or rejecting the current edge), without ever questioning the previous choices.

In Figure 7.15 a simple implementation is shown, based on the following data structure:

- $Edge[h].From$ = first endpoint of the h-th edge of G ($h = 1, \ldots, m$);

- $Edge[h].To$ = second endpoint of the h-th edge of G ($h = 1, \ldots, m$);

- $Edge[h].cost$ = cost of the h-th edge of G ($h = 1, \ldots, m$);

- k = number of edges chosen ($k = |T^*|$);

- $Tree[q].From$ = first endpoint of the q-th edge in T^* ($q = 1, \ldots, k$);

- $Tree[q].To$ = second endpoint of the q-th edge in T^* ($q = 1, \ldots, k$);

- $Tree[q].cost$ = cost of the q-th edge in T^* ($q = 1, \ldots, k$);

- $comp[i]$ = component containing vertex $i \in \{1, \ldots, n\}$.

The computational complexity of the sorting required at Step 1 is $O(m \log m)$, i.e., $O(m \log n)$ given that $\log m < \log n^2 = 2 \log n$. The update of the components requires $O(n)$ time and is repeated at most $n - 1$ times. It follows that the complexity of this

Kruskal's Algorithm;
  begin

1.   sort the edges of G according to the increasing costs, obtaining:
         $Edge[1].cost \leq Edge[2].cost \leq \ldots \leq Edge[m].cost$;
     $k := 0$;
     $h := 0$;

2.   **for** $i := 1$ **to** n **do** /* a vertex in each component */
        $comp[i] := i$;

3.   **while** $(k < n-1)$ **and** $(h < m)$ **do**
        begin /* consider the h-th edge */

4. $h := h + 1$;
          $i := Edge[h].From$;
          $j := Edge[h].To$;
          $C1 := comp[i]$;
          $C2 := comp[j]$;

5.   **if** $C1 \neq C2$ **then** /* select $[i, j]$ */
            begin

6. $k := k + 1$;
              $Tree[k] := Edge[h]$;

7.   **for** $q := 1$ **to** n **do** /* combine components $C1$ and $C2$ */
                 if $comp[q] = C2$ **then**
                    $comp[q] := C1$
          end
       end ;

8.   **if** $k \neq n-1$ **then** "graph G is not connected"
  end .

Figure 7.15: Kruskal's Algorithm

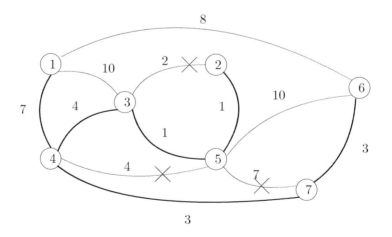

Figure 7.16: Application of Kruskal's Algorithm

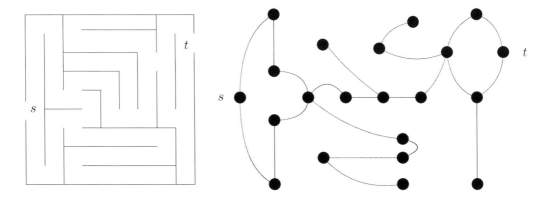

Figure 7.17: Mazes and paths on graphs

implementation is $O(m \log n + n^2)$, hence it is worse than that of Prim-Dijkstra's $O(n^2)$ algorithm. It is however possible to reduce the complexity of the method to $O(m \log n)$ by using an appropriate "union-find" data structure to represent the connected components of G_{T^*}.

An example of the application of Kruskal's algorithm is shown in Figure 7.16.

7.6 Paths

The problem of identifying a path in a graph has many practical applications. For instance, the problem of getting out of a maze can be interpreted as the problem of searching a path in a graph where the edges correspond to the pathways of the maze, see Figure 7.17. Another application consists in searching a path between two locations in a transport network in which the vertices are the intersections and the edges are the roads connecting them. In this latter case, each edge $e \in E$ typically has a cost c_e which indicates, for instance, the time needed to travel the corresponding road.

In the following we will study the most general case in which the considered graph is directed, hence the connecting cost may depend on the direction of the arcs. Note that each undirected graph can be transformed into a directed graph by replacing each edge $e = [i, j]$ with cost c_e with two arcs (i, j) and (j, i) with costs $c_{ij} := c_{ji} := c_e$.

As usual, we will indicate with $n := |V|$ and $m := |A|$ the number of vertices and arcs in the directed graph $G = (V, A)$.

PATHFINDING Algorithm ;
   begin
1.   **for** $j := 1$ **to** n **do**
        $pred[j] := 0;$
     $pred[s] := s;$
     $Q := \{s\}$;
2.   **while** $Q \neq \emptyset$ **do**
        begin /* process a reachable vertex $h \in Q$ */
3.   choose a vertex $h \in Q$ and set $Q := Q \setminus \{h\}$;
4.   **for each** $(h, j) \in \delta^+(h)$ **do**
5.   **if** $pred[j] = 0$ **then**
              begin
                $pred[j] := h$;
                $Q := Q \cup \{j\}$
              end
        end
   end .

Figure 7.18: PATHFINDING Algorithm for computing Γ_s^+

7.6.1   Reachability

The first simple problem we will consider is that of identifying the vertices which are reachable from a given vertex s. A possible algorithm for this problem is shown in Figure 7.18, and uses the following data structure:

- s = starting vertex (*source*)

- $pred[j]$ = vertex preceding j in a path from s to j ($= 0$ if vertex j is not reachable from s)

- Q = queue of the vertices reachable from s and not yet processed.

At each iteration of the **while-do** loop, we choose a vertex $h \in Q$ reachable from s and not yet processed, and we process it by identifying the new vertices j that can be directly reached from h. Note that each vertex j is inserted into the queue Q at most once, given that after the first insertion in Q we have $pred[j] \neq 0$. It follows that the **while-do** loop is performed at most once for the same vertex h, hence each arc (h, j) of the graph is considered at most once at Step 4. The procedure thus has complexity $O(|V| + |A|)$, provided that the graph is stored by means of the *forward-star* $\delta^+(v)$ of its vertices, as described in the related section of page 131.

In case we want to identify a path from s to a given vertex t, the algorithm can be stopped as soon as $pred[t] \neq 0$.

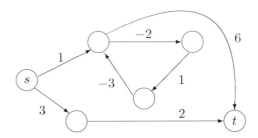

Figure 7.19: Graph with a negative cost cycle

7.6.2   Shortest Paths

Let us now consider the problem of identifying a *simple* path between two given vertices s and t of a directed graph $G = (V, A)$ with costs c_{ij} on the arcs. Contrary to what we may think, this optimization problem is *NP*-hard.

Theorem 7.6.1 *The problem of identifying a simple path with minimum cost in a graph is NP-hard.*

Proof: It is sufficient to observe that a simple path cannot have more than n arcs, and that a simple path has n arcs if and only if it is a Hamiltonian circuit. It follows that the problem of identifying a Hamiltonian circuit in G can be expressed as the problem of identifying the simple path from $s := 1$ to $t := 1$ containing the largest possible number of arcs, i.e., the simple path with minimum cost when each arc $(i, j) \in A$ is given a cost $c_{ij} := -1$. There exists hence an *NP*-complete problem (that of the Hamiltonian circuit) that can be reduced in polynomial time to the minimum-cost simple path problem, which is thus *NP*-hard.   □

There exist however important particular cases in which the problem is polynomial. As we will see, these cases occur whenever $c_{ij} \geq 0$ for all $(i, j) \in A$ (Dijkstra's Algorithm) or, more generally, when no negative-cost cycle exists (Floyd-Warshall's Algorithm).

ILP Model

For all $(i, j) \in A$, let $x_{ij} = 1$ if arc (i, j) is chosen in the path, $x_{ij} = 0$ otherwise. Assuming without loss of generality $s \neq t$, the simple path problem from s to t can be

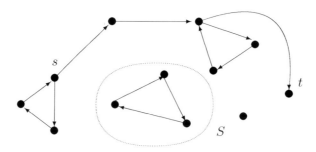

Figure 7.20: A possible solution of the model without *subtour elimination* constraints

formulated as:

$$\min \underbrace{\sum_{(i,j)\in A} c_{ij}x_{ij}}_{\text{path cost}} \tag{7.2a}$$

$$\underbrace{\sum_{(h,j)\in\delta^+(h)} x_{hj}}_{n.\ \text{leaving arcs}} - \underbrace{\sum_{(i,h)\in\delta^-(h)} x_{ih}}_{n.\ \text{entering arcs}} = \begin{cases} 1 & \text{if } h = s \\ -1 & \text{if } h = t \\ 0 & \text{for all } h \in \mathrm{V} \setminus \{s,t\} \end{cases} \tag{7.2b}$$

$$\underbrace{\sum_{(i,j)\,\in\,A(S)} x_{ij}}_{n.\ \text{arcs in } S} \le |S| - 1 \, , \, \forall \, S \subseteq V \, : \, S \neq \emptyset \tag{7.2c}$$

$$0 \le x_{ij} \le 1 \quad \text{integer} \, , \, (i,j) \, \in \, A \tag{7.2d}$$

The $2^n - 1$ constraints (7.2c) are called *subtour elimination* constraints, as they avoid the creation of "subtours" that use $|S|$ arcs with both endpoints in S. In the example of Figure 7.20 there exist 3 violated *subtour elimination* constraints, each of which corresponds to a subset S of the vertices covered by a subtour.

The solution of model (7.2a)-(7.2d) can be obtained by means of the *branch-and-bound* or *branch-and-cut* technique. Since the problem is *NP*-hard, it is indeed extremely unlikely that there exists a polynomial solution algorithm.

As mentioned before, there exist however particular cases that can be solved in polynomial time. Assume that G does not contain any cycle with negative or null cost. In such case, a solution as that in Figure 7.20 cannot be optimal, given that the elimination of the arcs of the subtours would result in a path of lower cost. A similar argument applies when there exist no cycles with negative cost, but cycles with null cost are admitted: if the solution of Figure 7.20 were optimal, it would then be easy to obtain a simple path with the same cost eliminating the subtours. It follows that constraints (7.2c) are redundant in this case, hence they can be eliminated from the model. In this case (but only in this case!) the integrality constraint becomes redundant as well, since the matrix of constraints (7.2b) coincides with the node-arc incidence matrix of G, hence it is totally unimodular.

In summary: if no negative cycles exist, the minimum-cost simple path problem can be formulated as a linear programming problem, hence it can be solved in polynomial time by using a polynomial algorithm for linear programming. Actually there exist specialized, and much more efficient, algorithms which exploit the specificity of the linear model and are able to efficiently solve problems with thousands of vertices and arcs.

In the following, we will therefore focus on the polynomial problem of identifying a simple path of minimum cost in a graph without negative cycles, omitting for the sake of brevity the adjective *simple* (made redundant by the hypothesis of absence of negative cycles).

7.6.3 Dijkstra's Algorithm

This algorithm requires $c_{ij} \geq 0$ for all $(i, j) \in A$ (a tighter hypothesis than that related to the absence of negative cycles), and allow us to compute in $O(n^2)$ time the shortest paths from a given vertex s to *all* other vertices t of the graph. These paths may be conveniently stored as an arborescence of root s, in which $pred[j]$ is the parent of vertex $j \in V$ in the arborescence. In this way, the shortest path from s to a given t can be re-created backwards, passing from t to $pred[t]$, from $pred[t]$ to $pred[pred[t]]$, and so on until we reach s.

Dijkstra's algorithm is based on the following property.

Theorem 7.6.2 *Assume that the cost L_i of the shortest paths from s to each vertex i belonging to a given set $S \subset V$ with $s \in S$ ($L_s := 0$) is known, and let $(v, h) := \arg\min\{L_i + c_{ij} : (i, j) \in \delta^+(S)\}$. If $c_{ij} \geq 0$ for all $(i, j) \in A$, then $L_v + c_{vh}$ is the cost of the shortest path from s to h.*

Proof: $L_v + c_{vh}$ is the cost of a path from s to h formed by the shortest path from s to v (with cost L_v) followed by arc (v, h). We have to prove that any other path P (say) from s to h has cost $c(P) \geq L_v + c_{vh}$. Let (i, j) be the first arc in $P \cap \delta^+(S)$, and as shown in Figure 7.21 partition P in $P_1 \cup \{(i, j)\} \cup P_2$, where P_1 and P_2 are two paths from s to i and from j to h, respectively. We have then

$$c(P) = \underbrace{c(P_1)}_{\geq L_i} + c_{ij} + \underbrace{c(P_2)}_{\geq 0} \geq L_i + c_{ij} \geq L_v + c_{vh}.$$

$\square$

The theorem immediately suggests the solution procedure shown in Figure 7.22, which is very similar to that of Prim-Dijkstra for the computation of the minimum spanning tree.

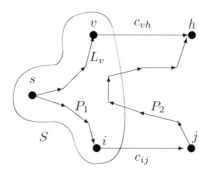

Figure 7.21: Proof of Theorem 7.6.2

Dijkstra's algorithm (1st version) ;
    begin
1. $S := \{s\}$;
      $L[s] := 0$;
      $pred[s] := s$;
2.   **while** $|S| \neq n$ **do**
        begin
3. $(v, h) := \arg\min\{L[i] + c[i,j] \; : \; (i,j) \in \delta^+(S)\}$;
4. $L[h] := L[v] + c[v,h]$;
          $pred[h] := v$;
          $S := S \cup \{h\}$
        end
    end .

Figure 7.22: Dijkstra's algorithm

At Step 3, if $\delta^+(S) = \emptyset$ the algorithm must be interrupted because all vertices $h \in V \setminus S$ cannot be reached from s.

An obvious implementation of Step 3 consists in scanning all arcs of the graph, rejecting those that do not belong to $\delta^+(S)$. In this way, we obtain a $O(m)$ complexity for a single execution of Step 3, and hence $O(n \cdot m)$ in total. As in the case of Prim-Dijkstra's algorithm for the computation of the minimum spanning tree, it is however possible to obtain a $O(n^2)$ complexity using a simple data structure.

Let us consider for the sake of simplicity the case of a complete directed graph G, stored by means of a $n \times n$ matrix of the costs $c[i,j]$ (with $c[i,j] = +\infty$ for the possible pairs (i,j) missing from the graph).

At each iteration and with respect to the current set S, for all $j \in V \setminus \{s\}$ let

- $flag[j] = \begin{cases} 1 & \text{if } j \in S \\ 0 & \text{otherwise} \end{cases}$

- $L[j] = \begin{cases} \text{cost of the min. path from } s \text{ to } j, & \text{if } j \in S \\ \min\{L[i] + c_{ij} : i \in S\}, & \text{if } j \notin S \end{cases}$

- $pred[j] = \begin{cases} \text{predecessor of } j \text{ in the minimum path from } s \text{ to} & \text{if } j \in S \\ j, & \\ \arg\min\{L[i] + c_{ij} : i \in S\}, & \text{if } j \notin S \end{cases}$

while, by definition, $flag[s] := 1$, $L[s] := 0$ and $pred[s] := s$.

Using these pieces of information, at Step 3 we can compute quantity $\min\{L[i] + c_{ij} : (i,j) \in \delta^+(S)\}$ as $\min\{L[j] : j \in V \setminus S\}$ in $O(n)$ time. In addition, after Step 3, we can update all values $L[j]$ and $pred[j]$ for all $j \notin S \cup \{h\}$, again in $O(n)$ time, given that we have:

$$\underbrace{\min\{L[i] + c[i,j] : i \in S \cup \{h\}\}}_{\text{new } L[j]} =$$
$$\min\{\underbrace{\min\{L[i] + c[i,j] : i \in S\}}_{\text{previous } L[j]}, L[h] + c[h,j]\}.$$

The algorithm is shown in Figure 7.23. Note that the algorithm is very similar to that of Prim-Dijkstra for the minimum spanning tree: the only difference is the updating formula for $L[j]$ at Step 6.

Each execution of Steps 4 and 6 needs $O(n)$ time, as does the initialization at Step 1. The complexity of the algorithm is hence $O(n^2)$.

As for Prim-Dijkstra's algorithm for the minimum spanning tree, in the case of sparse graphs with $m \ll n^2$, a $O(m \log n)$ implementation obtained using a priority queue Q to store vertices $j \notin S$ may be convenient.

Dijkstra's algorithm ($O(n^2)$ version);
   begin
1.   **for** $j := 1$ **to** n **do** /* initialization */
       begin
         $flag[j] := 0$;
         $pred[j] := s$;
         $L[j] := c[s, j]$
       end ;
2. $flag[s] := 1$; /* source vertex */
    $L[s] := 0$;
3.   **for** $k := 1$ **to** $n - 1$ **do** /* select a new arc */
       begin
4. $min := +\infty$; /* identify $h = \arg\min\{L[j] \, : \, j \notin S\}$ */
         for $j := 1$ **to** n **do**
           if $(flag[j] = 0)$ **and** $(L[j] < min)$ **then**
             begin
               $min := L[j]$;
               $h := j$
             end ;
5. $flag[h] := 1$; /* update $S := S \cup \{h\}$ */
6.   **for** $j := 1$ **to** n **do** /* update $L[j]$ and $pred[j]$ for all $j \notin S$ */
           if $(flag[j] = 0)$ **and** $(L[h] + c[h, j] < L[j])$ **then**
             begin
               $L[j] := L[h] + c[h, j]$;
               $pred[j] := h$
             end
       end
   end .

Figure 7.23: Dijkstra's $O(n^2)$ algorithm

Example

Consider the following example, in which costs c_{ij} are shown besides the corresponding arcs:

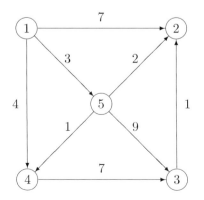

$[\, L[j], pred[j]\,]$

labels associated with vertices

By applying Dijkstra's algorithm from vertex $s = 1$, we obtain the following labels:

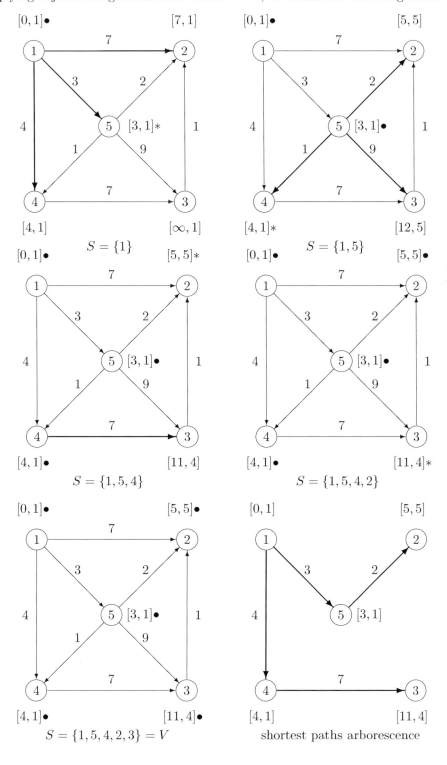

shortest paths arborescence

Floyd-Warshall's algorithm ;
  begin
1.   **for** $i := 1$ **to** n **do**
        for $j := 1$ **to** n **do**
          initialize $d[i,j] := c[i,j]$ and $pred[i,j] := i$;
2.   **for** $h := 1$ **to** n **do** /* triangular operation on h */
        begin
3.   **for** $i := 1$ **to** n **do**
            for $j := 1$ **to** n **do**
4.   **if** $d[i,h] + d[h,j] < d[i,j]$ **then**
                begin
                  $d[i,j] := d[i,h] + d[h,j]$;
                  $pred[i,j] := pred[h,j]$
                end ;
5.   **for** $i := 1$ **to** n **do**
            if $d[i,i] < 0$ **then** STOP "negative cycles"
        end
  end .

Figure 7.24: Floyd-Warshall 's algorithm

7.6.4   Floyd-Warshall's Algorithm

If we want to find the shortest paths from *all* vertices s to *all* other vertices t, we can apply Dijkstra's algorithm from each vertex s, with a $O(n^3)$ complexity. A different approach, still with complexity $O(n^3)$, is given by the Floyd-Warshall's algorithm. This algorithm can also be used in case of negative costs, and is able to identify possible negative-cost cycles. Obviously, if there exist negative cycles the correctness of the algorithm is not guaranteed.

The algorithm can be applied to a complete directed graph defined by the $n \times n$ matrix of costs $c[i,j]$. In output, the algorithm provides the following information, defined for all $i,j \in V$:

- $d[i,j]$ = cost of the shortest path from i to j

- $pred[i,j]$ = predecessor of j in the shortest path from i to j.

On output, each value $d[i,i]$ is the cost of the minimum cycle passing through vertex i: if $d[i,i] < 0$, then there exists a negative cycle (reconstructible from $pred[i,i]$) that passes from i. The algorithm is shown in Figure 7.24.

The double loop **for** i ... **for** j at Step 3 is called *triangular operation* associated with vertex h (set at Step 2). The update at Step 4 is based on the fact that the optimal path from i to j is composed of the optimal path from i to h, followed by the optimal path from h to j.

The correctness of the algorithm derives from the following theorem:

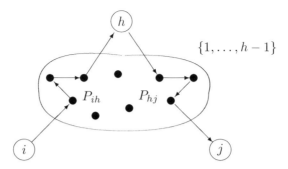

Figure 7.25: Proof of Theorem 7.6.3

Theorem 7.6.3 *Consider matrix* **d** *obtained after performing the triangular operation on vertices* $1, \ldots, h$. *In the absence of negative cycles, for all* $i, j \in V$, *value* $d[i, j]$ *is the cost of the shortest path from* i *to* j *in the subgraph induced by the set of vertices* $\{1, \ldots, h\} \cup \{i, j\}$.

Proof: The proof is by induction on h.

For $h = 0$, matrix **d** coincides with the initial matrix of the costs of the arcs, hence $d[i, j] = c[i, j]$ represents the cost of the shortest path from i to j in the subgraph induced by $\{i, j\}$, as claimed.

Let us now suppose that the property is verified at iteration $h - 1$. Consider a shortest path P_{ij} from i to j in the subgraph induced by $\{1, \ldots, h\} \cup \{i, j\}$, and let $c(P_{ij})$ be its cost. Two cases may occur:

(1) P_{ij} does not pass through vertex h: in this case $c(P_{ij}) = d[i, j]$ by the inductive hypothesis.

(2) P_{ij} passes through h: as shown in Figure 7.25, P_{ij} can be partitioned in this case in $P_{ih} \cup P_{hj}$, where P_{ih} is a shortest path from i to h in the subgraph induced by $\{1, \ldots, h-1\} \cup \{i, h\}$, and P_{hj} is a shortest path from h to j in the subgraph induced by $\{1, \ldots, h-1\} \cup \{h, j\}$. By the inductive hypothesis, we then have that $c(P_{ih}) = d[i, h]$ and $c(P_{hj}) = d[h, j]$, and hence $c(P_{ij}) = d[i, h] + d[h, j]$.

Therefore it holds that:

$$c(P_{ij}) \;=\; \min\{\underbrace{d[i, j]}_{\text{case (1)}}, \underbrace{d[i, h] + d[h, j]}_{\text{case (2)}}\}$$

thus, *after* the triangular operation on h, we have $d[i, j] = c(P_{ij})$, as requested. □

It is perhaps useful to observe that, in the presence of negative cycles, the shortest path P_{ij} defined in case (2) of the proof may not be simple. This situation would occur if both subpaths P_{ih} and P_{hj} passed through the same vertex $v < h$, hence P_{ij} would be formed by a path P'_{ij} from i to j passing through v but not through h, plus a cycle C' passing both through v and h; see Figure 7.25. In this case, we should have $c(P_{ij}) = c(P'_{ij}) + c(C') < c(P'_{ij})$ since case (1) of the proof did not occur, and hence $c(C') < 0$.

Example

We will now show the distance matrix **d** and the predecessor matrix **pred** computed by the Floyd-Warshall's algorithm for a simple example with $n = 4$ vertices related to the following graph

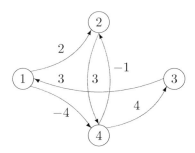

distance matrix

0	2	99	-4
99	0	99	3
3	99	0	99
99	-1	4	0

predecessor matrix

1	1	1	1
2	2	2	2
3	3	3	3
4	4	4	4

distance matrix

0	2	99	-4
99	0	99	3
3	(5)	0	(-1)
99	-1	4	0

predecessor matrix

1	1	1	1
2	2	2	2
3	(1)	3	(1)
4	4	4	4

distance matrix

0	2	99	-4
99	0	99	3
3	5	0	-1
(98)	-1	4	0

predecessor matrix

1	1	1	1
2	2	2	2
3	1	3	1
(2)	4	4	4

distance matrix

0	2	99	-4
99	0	99	3
3	5	0	-1
⑦	-1	4	0

predecessor matrix

1	1	1	1
2	2	2	2
3	1	3	1
③	4	4	4

distance matrix

0	⑤₋	⓪	-4
⑩	0	⑦	3
3	②₋	0	-1
7	-1	4	0

predecessor matrix

1	④	④	1
③	2	④	2
3	④	3	1
3	4	4	4

7.6.5 Project Planning

An important application of the shortest path problem can be found in the planning of complex projects.

Definition 7.6.1 *A project is a set of activities A_i $(i = 1, \ldots, m)$, each having a known duration $d_i \geq 0$. Among some activities, possible precedence relations of the type $A_i \prec A_j$ can be specified (meaning that activity A_j can begin only after the completion of activity A_i).*

One way to describe a project by means of a directed graph $G = (V, A)$ consists in associating an arc with each activity. The duration of each activity may be interpreted as a cost associated with the corresponding arc. The arcs are arranged in the graph such that the following property is verified: $A_i \prec A_j$ if and only if G contains a path that includes both arcs associated with A_i and A_j.

Consider, for instance, the following project:

activity : A, B, C, D, E ;   precedence : $A \prec B, A \prec C, B \prec D, C \prec D, B \prec E$.

A first representation of this project can be obtained introducing a *dummy activity*, having zero duration, for each of the 5 precedence relations specified. We obtain thus the following directed graph:

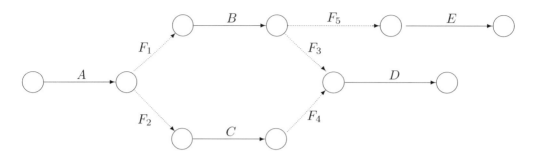

A more concise representation, with a single dummy activity, can be obtained by contracting the dummy activities F_1, F_2, F_4 and F_5:

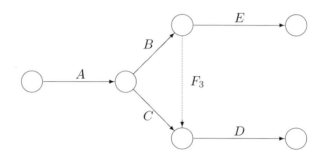

When simplifying the graph, care should be taken not to introduce spurious precedences. For instance, when contracting by mistake activity F_3 instead of F_4, we would obtain graph

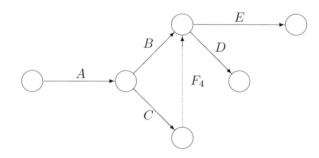

in which we have introduced the precedence relation $C \prec E$ that did not exist in the initial project.

In the following, we will not distinguish between an activity A_h of the project and the arc (i, j) that represents it, and we will indicate with d_{ij} its duration ($d_{ij} = 0$ for dummy activities).

Observe that graph G associated with a project does not contain cycles, given that, in this case, there would exists a logical inconsistency of the type $A_i \prec A_j \prec \ldots \prec A_k \prec A_i$.

Vertices v of G may be interpreted as "events" corresponding to the end of all activities associated with arcs $(i,v) \in \delta^-(v)$ entering the vertex, and hence to the possible beginning of activities $(v,j) \in \delta^+(v)$.

The techniques to be described require that the graph G representing the project has a single *initial* vertex s (i.e., such that $\delta^-(s) = \emptyset$) and a single final vertex t (i.e., such that $\delta^+(t) = 0$), and does not contain multiple arcs. By introducing dummy arcs and/or nodes, it is always possible to obtain such a graph, proceeding as illustrated in the figure below:

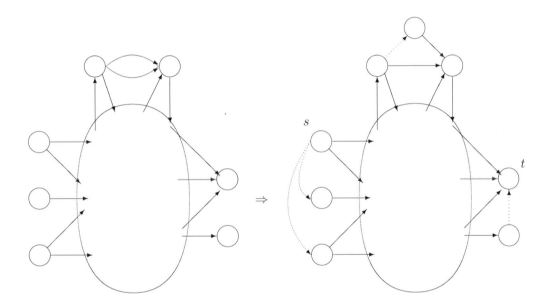

It is also necessary to label the vertices in *topological order*, i.e., to make sure that $i < j$ holds for each arc $(i,j) \in A$. Such a labeling can be obtained by:

1. assigning number 1 to any vertex v with $\delta^-(v) = \emptyset$;

2. eliminating from the graph vertex v with its arcs;

3. repeating the procedure to assign, in sequence, numbers $2, 3, \ldots, n$ to the remaining vertices.

Planning a project means scheduling its activities so as to minimize the *overall duration* of the project, defined as the end of the last activity planned. Indicating with t_i the time instant when the event associated with vertex i occurs, i.e.

$$t_i = \text{moment of the beginning of activities } (i,j) \in \delta^+(i),$$

CPM algorithm ;
  begin
1. $TMIN[1] := 0$;
    sort the vertices topologically;
2.   **for** $h := 2$ **to** n **do**
3. $TMIN[h] := \max\{TMIN[i] + d[i,h] \; : \; (i,h) \in \delta^-(h)\}$;
4. $TMAX[n] := TMIN[n]$; /* minimum duration of the project */
5.   **for** $h := n - 1$ **downto** 1 **do**
        $TMAX[h] := \min\{TMAX[j] - d[h,j] \; : \; (h,j) \in \delta^+(h)\}$
  end .

Figure 7.26: CPM Algorithm

we easily obtain the following linear programming model:

$$\min \; t_n \qquad\qquad\qquad \text{(duration of the project)}$$

$$t_1 = 0 \qquad\qquad\qquad \text{(beginning of the first activity)}$$

$$t_j \geq \underbrace{t_i + d_{ij}}_{\text{end of activity } (i,j)} \;\;, \text{ for all } (i,j) \in A \qquad\qquad \text{(precedences)}$$

$$t_i \geq 0 \; , \; i = 1, \ldots, n.$$

It is possible to show that the minimum duration of the project coincides with the length of the *longest* path from 1 to n in G, i.e., with the length of the shortest path with costs $c_{ij} := - d_{ij}$ on the arcs. Since the graph is acyclic, there cannot exist negative circuits, hence this problem can be solved using, for instance, Floyd-Warshall's $O(n^3)$ algorithm.

In practice, it is better to use a more efficient algorithm, with complexity $O(m)$, that exploits the acyclicity of G. This algorithm is generally known as the *CPM technique* (Critical Path Method), or as the *PERT technique* (Project Evaluation and Review Technique) if the durations of the activities are random variables; see Figure 7.26.

The algorithm associates each vertex $h \in V$ with an *earliest time* $TMIN_h$ before which the corresponding event cannot occur. We hence have that $TMIN_n$ is the minimum duration of the project. For each vertex h a *latest time* $TMAX_h$ is also computed, corresponding to the latest time in which the event can occur without compromising the minimum duration of the project.

Starting from values $TMIN$ and $TMAX$ computed in such a way, it is possible to define for each activity $(i,j) \in A$:

- the earliest starting time $(:= TMIN_i)$;

- the latest starting time $(:= TMAX_i)$, after which there is a delay with respect to the minimum duration of the entire project;

- the maximum *float* allowed if the remaining activities do not suffer delays: $TMAX_j - TMIN_i - d_{ij}$.

Activities (i, j) with zero float are said to be *critical*, and satisfy $TMIN_i = TMAX_i$, $TMIN_j = TMAX_j$, and $TMIN_i + d_{ij} = TMAX_j$.

A path from s to t in G is said to be *critical* if all its arcs are critical.

Each project has at least one critical path corresponding to a critical sequence of activities, delaying any of which inevitably causes a delay in the completion time of the project.

Example

In the following, a graph related to a simple project is shown, with durations d_{ij} besides the arcs and labels $[TMIN_h, TMAX_h]$ besides the vertices.

The arcs of the critical path (unique in this example) are shown in bold.

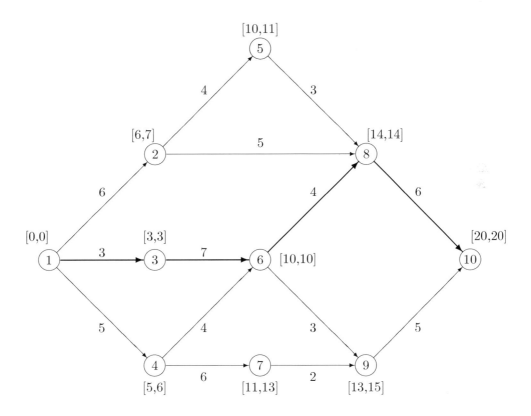

A time representation of the project, called *Gantt chart*, is the following (we assume that each activity (i, j) starts at $TMIN_i$ time):

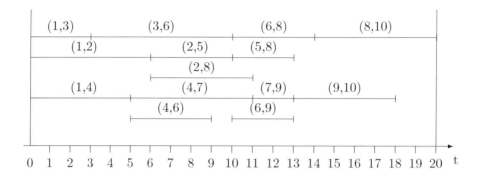

7.7   Flow Problems

Some important applications of graph theory consider systems for the distribution of a certain product (for instance: water, methane, telephone calls, etc.) from one or more production points to one or more points of use. These problems can be described by particular directed graphs, called flow networks.

A *flow network* is a directed graph $G = (V, A)$ in which a *capacity* $k_{ij} \geq 0$, and possibly a cost c_{ij}, are associated with each arc $(i, j) \in A$. In addition, two vertices s and t are specified, called *source* vertex and *sink* vertex, respectively. Typically, but not always, we have $\delta^-(s) = \delta^+(t) = 0$, i.e., the network does not have arcs entering into s nor arcs leaving from t.

Definition 7.7.1 *Given a flow network, a* feasible flow *from s to t is a function* $\mathbf{x} : A \to \Re$ *such that:*

$$0 \leq x_{ij} \leq k_{ij} , \ (i, j) \in A \tag{7.3}$$

$$\underbrace{\sum_{(h,j) \,\in\, \delta^+(h)} x_{hj}}_{\text{flow leaving from } h} - \underbrace{\sum_{(i,h) \,\in\, \delta^-(h)} x_{ih}}_{\text{flow entering into } h} = 0 , \ h \in V \setminus \{s, t\}. \tag{7.4}$$

Conditions (7.3) express the fact that flow x_{ij} on each arc (i, j) cannot be negative, nor exceed capacity k_{ij}. Conditions (7.4) impose that the flow entering into each vertex $h \neq s, t$ must be equal to the flow leaving it.

An example of feasible flow on a network is shown in Figure 7.27.

The most common network optimization problem consists in sending the maximum flow from s to t, and corresponds to the following linear programming problem:

$$\mathbf{MAX - FLOW}: \ \max\{\varphi_0 := \sum_{(s,j)\in\delta^+(s)} x_{sj} - \sum_{(i,s)\in\delta^-(s)} x_{is} : \text{constraints } (7.3) - (7.4)\}.$$

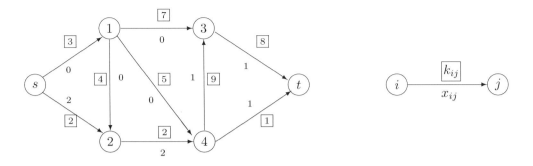

Figure 7.27: Example of feasible flow

A more general problem consists in identifying a minimum-cost flow from s to t, and is defined by the linear programming model that follows:

MIN – COST FLOW : $\quad \min\{ \sum_{(i,j) \,\in\, A} c_{ij}x_{ij} \; : \; \text{constraints } (7.3) - (7.4)\},$

where typically $c_{sj} \ll 0$ for arcs $(s,j) \in \delta^+(s)$, and $c_{is} \gg 0$ for arcs $(i,s) \in \delta^-(s)$.

Note that the constraint matrix of these linear programming problems is totally unimodular, given that constraint matrix (7.4) is composed by the rows of the node-arc incidence matrix of graph G associated with vertices $h \neq s, t$. It follows that a flow problem always has an optimal integer solution when capacities k_{ij} are all integer.

The network model can often be used even in the case of apparently more difficult problems. Let us see same simple examples.

If there is a network with more than one source vertex and/or more than one sink vertex, it is possible to reduce to the regular case by creating a dummy source s^* from which arcs with infinite capacity go towards the real sources, and a dummy sink vertex t^* to which arcs with infinite capacity come from the real sinks. Note that the transformation cannot be used if the "goods" transported from the various sources to the various sinks are different from each other (leading to an *NP*-hard problem known as the *multi-commodity flow* problem).

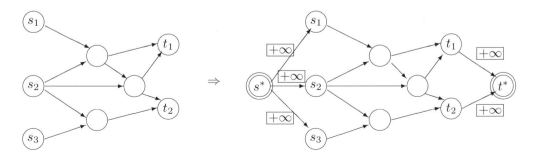

If some vertices $v \neq s, t$ have a capacity k_v, corresponding to a constraint $\sum_{(i,v) \in \delta^-(v)} x_{iv} = \sum_{(v,j) \in \delta^+(v)} x_{vj} \leq k_v$, it is possible to: (1) replace v with two new vertices, say v^+ and v^-; (2) replace arcs (i, v) entering into v with (i, v^-), and arcs (v, j) leaving from v with (v^+, j); and finally (3) introduce an arc (v^-, v^+) with capacity k_v.

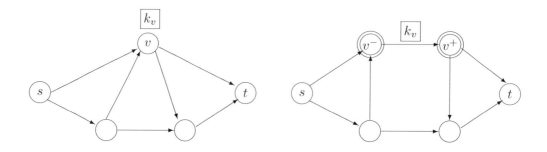

Finally, if the starting graph is undirected (or contains some undirected edges) then it is often possible to replace each edge $[i, j]$ with two arcs (i, j) and (j, i) with the same capacity and cost. This construction may however fall short if the problem considered is the MIN-COST FLOW problem and edge $[i, j]$ has a negative cost: the new pair (i, j), (j, i) induces in this case a negative cycle along which it is possible to make the flow "circulate" by erroneously decreasing the overall cost.

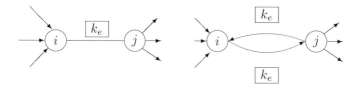

7.7.1   Fundamental Properties

We will now prove some fundamental properties of network flows.

Definition 7.7.2 *We call* cut *a partition $(S, V \setminus S)$ of the set of vertices such that $s \in S$ and $t \in V \setminus S$.*

Definition 7.7.3 *Given a feasible flow $\mathbf{x}$, we call* flow through a cut $(S, V \setminus S)$ *the quantity:*

$$\varphi(\mathrm{S}) := \underbrace{\sum_{(i,j) \, \in \, \delta^+(S)} x_{ij}}_{\text{flow leaving from } S} - \underbrace{\sum_{(i,j) \, \in \, \delta^-(S)} x_{ij}}_{\text{flow entering into } S}.$$

Quantity

$$\varphi_0 := \varphi(\{s\}) = \sum_{(s,j)\,\in\,\delta^+(s)} x_{sj} - \sum_{(i,s)\,\in\,\delta^-(s)} x_{is}$$

is said to be the value *of flow* **x**.

Definition 7.7.4 *The* capacity of the cut $(S, V \setminus S)$ *is the quantity:*

$$k(S) := \sum_{(i,j)\in\delta^+(S)} k_{ij}.$$

Note that arcs $(i,j) \in \delta^-(S)$ *entering* into S do not contribute to the capacity of the cut. The following theorems apply:

Theorem 7.7.1 *Let* **x** *be a feasible flow. For every cut* $(S, V \setminus S)$ *one has* $\varphi(S) = \varphi_0$, *i.e., the flow through every cut is a constant.*

Proof: Consider an arbitrary cut $(S, V \setminus S)$. Then:

$$\varphi_0 := \varphi(\{s\}) := \sum_{(s,j)\in\delta^+(s)} x_{sj} - \sum_{(i,s)\in\delta^-(s)} x_{is} = \sum_{h\,\in\, S} \underbrace{\left[\sum_{(h,j)\in\delta^+(h)} x_{hj} - \sum_{(i,h)\in\delta^-(h)} x_{ih} \right]}_{=0 \text{ for all } h \,\neq\, s}$$

$$= \sum_{h\,\in\, S}\sum_{(h,j)\in\delta^+(h)} x_{hj} - \sum_{h\,\in\, S}\sum_{(i,h)\in\delta^-(h)} x_{ih}$$

$$= \left[\underbrace{\sum_{(i,j)\in A(S)} x_{ij}}_{=} + \sum_{(i,j)\in\delta^+(S)} x_{ij} \right] - \left[\underbrace{\sum_{(i,j)\in A(S)} x_{ij}}_{=} + \sum_{(i,j)\in\delta^-(S)} x_{ij} \right] =: \varphi(S).$$

□

Theorem 7.7.2 *For every feasible flow* **x** *and every cut* $(S, V \setminus S)$, *one has*

$$\varphi(S) \leq k(S).$$

Proof:

$$\varphi(S) := \sum_{\substack{(i,j)\in\delta^+(S) \\ \underbrace{}_{\leq k_{ij}}}} x_{ij} - \sum_{\substack{(i,j)\in\delta^-(S) \\ \underbrace{}_{\geq 0}}} x_{ij} \leq \sum_{(i,j)\in\delta^+(S)} k_{ij} =: k(S).$$

□

The two previous theorems imply a "weak" duality relation: the value φ_0^* of the maximum flow cannot exceed the capacity $k(S^*)$ of the minimum-capacity cut $(S^*, V\setminus S^*)$. Actually, we will constructively prove that φ_0^* is always *equal* to the minimum capacity $k(S^*)$, thus obtaining the following "strong" duality relation:

MAX-FLOW/MIN-CUT:*A feasible flow* **x** *is optimal for the MAX-FLOW problem if and only if there exists a cut* $(S^*, V\setminus S^*)$ *with* $\varphi(S^*) = k(S^*)$. *In this case,* $(S^*, V\setminus S^*)$ *is a cut of minimum capacity in the given network.*

$$\varphi_0^*$$

$$\longrightarrow \max_{\mathbf{x}}\{\varphi_0\} \quad = \quad \min_S\{k(S)\} \longleftarrow$$

To constructively prove the theorem, we will introduce the concept of *residual* network, which considers the possible flow *modifications* with respect to a given current flow **x**.

Definition 7.7.5 *Let* **x** *be a feasible flow. An arc* (i,j) *is said to be saturated if* $x_{ij} = k_{ij}$, *empty if* $x_{ij} = 0$.

Definition 7.7.6 *Let* **x** *be a feasible flow. The residual network* $\overline{G} = (V, \overline{A})$ *associated with* **x** *is obtained from the original network* $G = (V, A)$ *by replacing each arc* $(i,j) \in A$ *with two arcs:*

- *a forward arc* (i,j) *of residual capacity* $\overline{k}_{ij} := k_{ij} - x_{ij} \geq 0$
- *a backward arc* (j,i) *of residual capacity* $\overline{k}_{ji} := x_{ij} \geq 0$

and then eliminating all arcs with zero residual capacity (see Figure 7.28).

The residual network therefore takes into account the possibility to increase the flow on unsaturated arcs (those that give rise to forward arcs with non-zero residual capacity), and decrease it on non-empty arcs (those that give rise to backward arcs with non-zero residual capacity). Note that an incremental flow $\overline{x}_{ji}$ on a *backward* arc $(j,i) \in \overline{A}$ can be interpreted as a *decrease* of the current flow x_{ij} on the arc $(i,j) \in A$ that originated it.

It is easy to see that a path from s to t in the residual network identifies the possibility to increase value φ_0 of the current flow. Such a path is called *augmenting path*, and corresponds in the original network to a sequence of forward arcs (along which it is possible to increase the flow) and backward arcs (along which it is possible to decrease the flow); see the example in Figure 7.29. The following property holds:

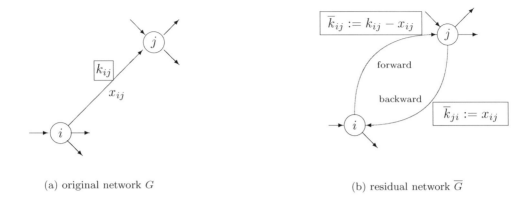

(a) original network G (b) residual network $\overline{G}$

Figure 7.28: Residual network: arcs (u, v) with $\overline{k}_{uv} = 0$ must be eliminated

Theorem 7.7.3 *A feasible flow* **x** *is optimal for the MAX-FLOW problem if and only if vertex t is not reachable from vertex s in the residual network $\overline{G} = (V, \overline{A})$ associated with* **x**.

Proof: Let φ_0 be the value of flow **x**. If t is reachable from s in $\overline{G}$, there exists then an augmenting path P from s to t in $\overline{G}$. Setting $\delta := \min\{\overline{k}_{uv} : (u, v) \in P\} > 0$, for all $(u, v) \in P$ it is possible to update $x_{uv} := x_{uv} + \delta$ if (u, v) is a forward arc; $x_{vu} := x_{vu} - \delta$ if (u, v) is a backward arc. It is easy to verify that the new vector **x** defines a feasible flow of value $\varphi_0 + \delta$ in the original network, which proves that the starting solution **x** was not optimal for the MAX-FLOW problem.

Let us now assume that t is not reachable from s in $\overline{G}$. There hence exists a cut $(S^*, V \backslash S^*)$ in the *residual* network $\overline{G}$ such that $\delta_{\overline{G}}^{\pm}(S^*) = \emptyset$. By the definition of residual network we have that, in the *original* network G:

- each arc $(i, j) \in \delta_G^+(S^*)$ is saturated

- each arc $(i, j) \in \delta_G^-(S^*)$ is empty.

It follows that

$$\varphi(S^*) := \underbrace{\sum_{(i,j) \in \delta_G^+(S^*)} x_{ij}}_{\text{all saturated}} - \underbrace{\sum_{(i,j) \in \delta_G^-(S^*)} x_{ij}}_{\text{all empty}} = \sum_{(i,j) \in \delta_G^+(S^*)} k_{ij} =: k(S^*)$$

hence the optimality of **x** derives from Theorem 7.7.2, which also guarantees that $(S^*, V \backslash S^*)$ is a cut of minimum capacity in the original network. $\qquad\square$

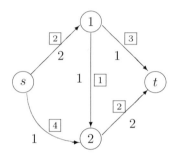

(a) flow **x** of value 3

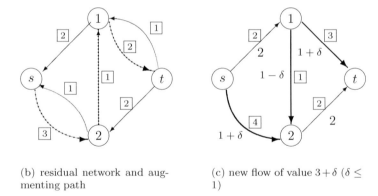

(b) residual network and augmenting path

(c) new flow of value $3+\delta$ ($\delta \leq$
1)

Figure 7.29: Increase of the flow

Ford-Fulkerson algorithm (1st version) ;
   begin
1.   x := 0; /* initialization */
       φ_0 := 0;
       optimal := **false**;
2.   **repeat**
3.   create the residual network $\overline{G} = (V, \overline{A})$ associated with x ;
4.   identify, if there exists, a path P from s to t in $\overline{G}$;
5.   **if** P does not exist **then**
           optimal := **true**
        else
          begin
6. $\delta := \min\{\overline{k}[u, v] \,:\, (u, v) \in P\}$;
            $\varphi_0 := \varphi_0 + \delta$;
7.   **for each** $(u, v) \in P$ **do**
              if (u, v) is a forward arc **then**
                $x[u, v] := x[u, v] + \delta$
              else
                $x[v, u] := x[v, u] - \delta$
         end
     until *optimal*= **true**
  end .

Figure 7.30: Ford-Fulkerson algorithm for the MAX-FLOW problem

7.7.2 Ford-Fulkerson Algorithm for the MAX-FLOW problem

Theorem 7.7.3 suggests a simple solution procedure for the MAX-FLOW problem, shown in Figure 7.30.

At Step 4, path P can be identified using the PATHFINDING algorithm described in the related section of page 148.

Since $\delta > 0$ at Step 6, at each iteration of the **repeat-until** loop the value φ_0 of the current flow monotonically increases, ensuring thus the asymptotic convergence of the procedure to the optimal value φ_0^*. In practice, capacities k_{ij} are often represented by integer numbers. In this case, it is easy to verify that at each iteration vectors x and $\overline{k}$ are as well integer, hence $\delta \geq 1$ and the algorithm converges after, at most, φ_0^* iterations. Note that $\varphi_0^* \leq k(\{s\}) = O(mk_{\max})$, where $k_{\max} := \max\{k_{ij} \,:\, (i, j) \in A\}$, and that each iteration of the **repeat-until** loop needs $O(m)$ time. In case of integer capacities the complexity of Ford-Fulkerson algorithm is hence $O(m^2 k_{\max})$, i.e., *non*-polynomial on the size of the instance that is equal to $O(m \log k_{\max})$.

There are indeed situations in which the algorithm can be extremely inefficient. Consider for instance the network of Figure 7.31, where M is a very big positive number. By systematically choosing the augmenting path that involves forward arc $(2, 3)$ or the corresponding backward arc $(3, 2)$, it is possible to increase the current flow of only $\delta = 1$ units at each iteration. Therefore $2M$ iterations are needed to identify the optimal flow.

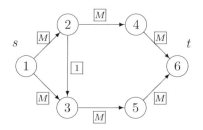

(a) original network

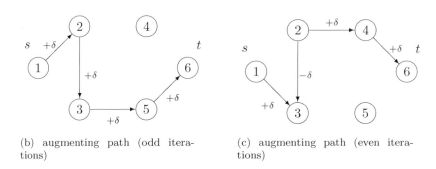

(b) augmenting path (odd iterations)

(c) augmenting path (even iterations)

Figure 7.31: Increase of the flow

There exist however simple modifications of the Ford-Fulkerson algorithm ensuring it a
polynomial complexity.

In Figure 7.32, a more detailed description of Ford-Fulkerson algorithm is shown, in
which instead of explicitly creating the residual network, the PATHFINDING algorithm
is modified so as to consider the forward and backward arcs created by the arcs of the
original network.

At each iteration of the loop **repeat-until**, each node $j \in V$ is labeled with two values
$[\varepsilon[j], \pm pred[j]]$, where $pred[j]$ is the vertex that precedes j in the augmenting path from
s to j, and $\varepsilon[j]$ is the amount of "residual flow" which can be sent to j through that
augmenting path.

The sign "+" associated with $pred[j]$ is used to indicate that vertex j can be reached, in
the residual network, through the *forward* arc associated with $(pred[j], j) \in A$; a "$-$"
sign on the other hand indicates that j can be reached, on the same residual network,
through the *backward* arc associated with the original arc $(j, pred[j]) \in A$.

Example

Let us consider the following flow network:

Ford-Fulkerson algorithm ;
   begin
1.   **for each** $(i, j) \in A$ **do** $x[i, j] := 0$; /* initial null flow */
     $\varphi_0 := 0$;
2.   **repeat** /* MAIN LOOP */
3.   **for each** $j := 1$ **to** n **do** $pred[j] := 0$;
       $\varepsilon[s] := +\infty$; $pred[s] := s$;
       $Q := \{s\}$; /* queue of labeled yet unprocessed vertices */
       while $(Q \neq \emptyset)$ **and** $(pred[t] = 0)$ **do**
         begin
4.   pop a vertex $h \in Q$ from the queue (e.g., $h := \min\{i \; : \; i \in Q\}$) ;
           $Q := Q \setminus \{h\}$;
5.   **for each** $(h, j) \in \delta_G^+(h) \; : \; x[h, j] < k[h, j]$ **do** /* unsaturated forward arcs */
             if $pred[j] = 0$ **then**
               begin
                 $\varepsilon[j] := \min\{\varepsilon[h], k[h, j] - x[h, j]\}$; $pred[j] := h$;
                 $Q := Q \cup \{j\}$
               end ;
6.   **for each** $(i, h) \in \delta_G^-(h) \; : \; x[i, h] > 0$ **do** /* non-empty backward arcs */
             if $pred[i] = 0$ **then**
                begin
                 $\varepsilon[i] := \min\{\varepsilon[h], x[i, h]\}$; $pred[i] := -h$;
                 $Q := Q \cup \{i\}$
               end
         end ;
7.   **if** $pred[t] \neq 0$ **then** /* augmenting path found */
        begin
8. $\delta := \varepsilon[t]$; $\varphi_0 := \varphi_0 + \delta$;
         $j := t$;
9.   **while** $j \neq s$ **do**
           begin /* backward reconstruction of the augmenting path */
             $i := pred[j]$;
             if $i > 0$
               then $x[i, j] := x[i, j] + \delta$
               else $x[j, -i] := x[j, -i] - \delta$;
             $j := |i|$
           end
        end
10.   **until** $pred[t] = 0$;
11.   "the current flow **x** is optimal, and a cut $(S^*, V \setminus S^*)$ of minimum capacity
     is the one associated with $S^* := \{j \in V \; : \; pred[j] \neq 0\}$"
   end .

Figure 7.32: Ford-Fulkerson algorithm implementation.

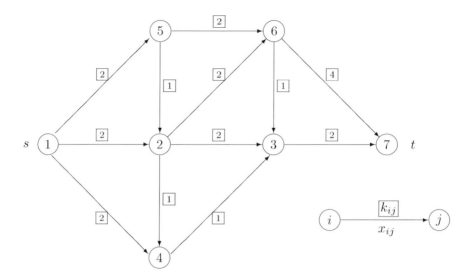

The initial flow **x** is zero. Then, labels $[\varepsilon[j], \pm pred[j]]$ are computed for each vertex of the network.

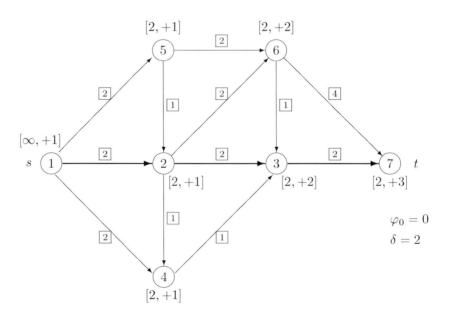

It is possible to send $\delta = 2$ units of residual flow through the augmenting path highlighted in bold in the previous figure. We thus obtain a new flow of value $\varphi_0 = 2$, in which the flow on arcs $(1, 2)$, $(2, 3)$, $(3, 7)$ is 2. This solution is represented in the next figure. Executing the labeling algorithm from this new solution, we get:

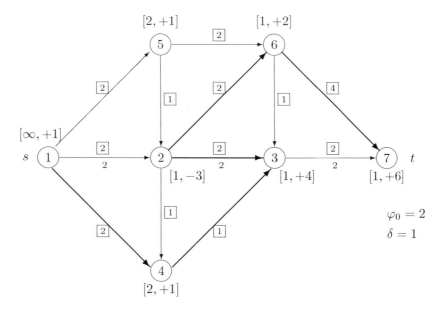

Note label $[1, -3]$ associated with vertex 2, which indicates that, in the residual network, a unit of residual flow can be sent to vertex 2 through the backward arc $(3, 2)$, associated with arc $(2, 3)$ in the original network. The current flow can be augmented again, incrementing by $\delta = 1$ the flow on arcs $(1, 4)$, $(4, 3)$, $(2, 6)$ and $(6, 7)$, and decrementing it by $\delta = 1$ on the arc $(2, 3)$. A new feasible flow of value $\varphi_0 = 3$ is thus obtained, which is represented in the next figure. Executing the labeling algorithm again from this new solution, we get:

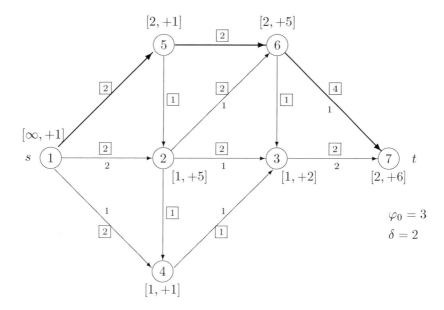

Incrementing by $\delta = 2$ the flow on the augmenting path $P = \{(1,5),(5,6),(6,7)\}$ we get a new solution of value $\varphi_0 = 5$, from which we can compute the following labels:

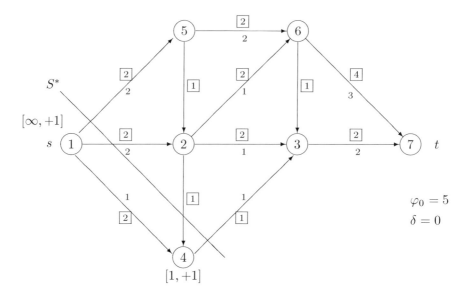

Unlabeled vertices j have $pred[j] = 0$, and are not reachable from s in the residual network. Given that $pred[t] = 0$, the current flow is optimal and a cut $(S^*, V \setminus S^*)$ of minimum capacity, shown in the figure, is associated with $S^* := \{1,4\}$. Note that $\varphi_0 = \varphi(S^*) = k(S^*) = 5$, and that all arcs $(i,j) \in \delta_G^+(S^*)$ are saturated, while all arcs $(i,j) \in \delta_G^-(S^*)$ are empty.   □

Chapter 8

Some NP-hard Problems

In the following, we will introduce some of the most common *NP*-hard problems.

8.0.3 Knapsack Problem

Let a set of items $\{1, \ldots, n\}$ be given, each with a *profit* p_j and a *weight* w_j, and a container of *capacity* W.

Without loss of generality, we can assume that all values p_j, w_j and W are positive integers, with $w_j < W$ for all $j = 1, \ldots, n$, and $\sum_{j=1}^{n} w_j > W$.

The *Knapsack Problem* (KP) consists in selecting an item subset with maximum profit to load into the container. This is a fundamental problem that arises whenever we want to optimally load any container (truck, cargo ship, etc...).

Introducing the decision variables

$$
x_j = \begin{cases} 1 & \text{if item } j \text{ is selected} \\ 0 & \text{otherwise} \end{cases}
$$

we get the ILP model:

$$z^* := \max \sum_{j=1}^{n} p_j x_j \tag{8.1a}$$

$$\sum_{j=1}^{n} w_j x_j \leq W \tag{8.1b}$$

$$0 \leq x_j \leq 1 \text{ integer }, \ j \in \{1, \ldots, n\}. \tag{8.1c}$$

This model is usually solved by *branch-and-bound*. It is easy to see that a fractional solution of model (8.1a)-(8.1c) can be interpreted as partially selecting some items; for example, $x_j = 0.3$ means selecting 30% of item j. This interpretation suggests an algorithm (due to G. Dantzig) to solve the continuous relaxation of model (8.1a)-(8.1c) by:

1. split each item j into w_j *subitems* of unitary weight and profit p_j/w_j;

2. fully load the container selecting the subitems of maximal profit p_j/w_j.

In other words, the algorithm sorts the items $j \in \{1, \ldots, n\}$ by non-increasing relative profit

$$\frac{p_1}{w_1} \geq \frac{p_2}{w_2} \geq \ldots \geq \frac{p_n}{w_n},$$

and finds the *critical item* $s \in \{1, \ldots, n\}$, defined as the one with the property:

$$\sum_{j=1}^{s-1} w_j < W \leq \sum_{j=1}^{s} w_j.$$

The optimal solution $\mathbf{x}^*$ of the continuous relaxation can then be obtained by:

1. fully selecting the first $s - 1$ items $\Rightarrow x_1^* = \ldots = x_{s-1}^* := 1$;

2. partially selecting the critical item $s \Rightarrow x_s^* := \left(W - \sum_{j=1}^{s-1} w_j \right) / w_s$;

3. discarding the following items $\Rightarrow x_{s+1}^* = \ldots = x_n^* := 0$.

This way it is possible to solve the continuous relaxation in $O(n \log n)$ time, or even in $O(n)$ time by using a so-called *partial sorting* algorithm.

If capacity W is a relatively small integer value, it is also possible to solve the knapsack problem by *dynamic programming*, using the algorithm shown in Figure 8.1 which can be described as follows.

For each $j = 0, \ldots, n$ and $K = 0, \ldots, W$, let

$$z[K, j] = \begin{cases} \text{maximum profit that can be obtained with the subset of items } 1, \ldots, j \\ \text{on a container with capacity } K. \end{cases}$$

The optimal value z^* of the given KP is thus $z[W, n]$.

At Step 1, column $j = 0$ of matrix z is initialized. Then items $j = 1, \ldots, n$ are iteratively processed, and for each capacity $K = 0, \ldots, W$ the algorithm decides whether the optimal loading of the container of capacity K with the first j items coincides with the one using the first $j - 1$ items, i.e., item j is not selected (Steps 5 and 9), or whether it is convenient to select item j (with profit $p[j]$) and thus optimally load with the remaining items $1, \ldots, j - 1$ the residual capacity $K - w[j]$ (with profit $z[K - w[j], j - 1]$; see Step 8). By Step 10, all $(W + 1) \times (n + 1)$ elements of matrix z are computed, in $O(nW)$ time. In practice, we have solved *all* knapsack problems defined by the given items and

PD-KP Algorithm;

  begin

1.   **for** $K := 0$ **to** W **do** /* initialization for $j = 0$ */

      $z[K, 0] := 0$;

2.   **for** $j := 1$ **to** n **do** /* define column j of the matrix **z** */

      begin

3. $weight := w[j]$;

4.   **for** $K := 0$ **to** $weight - 1$ **do** /* insufficient capacity K for item j */

5. $z[K, j] := z[K, j - 1]$;

6.   **for** $K := weight$ **to** W **do** /* enough capacity K for item j */

7.   **if** $p[j] + z[K - weight, j - 1] > z[K, j - 1]$ **then**

8. $z[K, j] := p[j] + z[K - weight, j - 1]$ /* it is better to select j */

          else

9. $z[K, j] := z[K, j - 1]$ /* it is better not to select j */

     end ;

10. $z^* := z[W, n]$; /* optimal solution value */

     $ResidualCap := W$;

11.   **for** $j := n$ **downto** 1 **do** /* define the optimal solution $x^*[j]$ */

12.   **if** $z[ResidualCap, j] = z[ResidualCap, j - 1]$ **then**

13. $x^*[j] := 0$

     else

       begin

14. $x^*[j] = 1$;

15. $ResidualCap := ResidualCap - w[j]$

       end

  end .

Figure 8.1: Dynamic programming algorithm for KP

by capacities $K = 1, \ldots, W$. As we are interested in the optimal solution for capacity W, at Step 10 we iterate backward over the columns $j = n, n - 1, \ldots, 1$, starting from entry $z[W, n]$ of the matrix, and reconstruct the optimal choices $x_n^*, x_{n-1}^*, \ldots, x_1^*$. This reconstruction has complexity $O(n)$.

Algorithm PD-KP has thus complexity $O(nW)$, *non* polynomial in the size $O(n \log W)$ of the instance. In addition, it requires a matrix of size $(W + 1) \times (n + 1)$ to store the intermediate values of the computation. For large values of W, this algorithm is thus not competitive with *branch-and-bound* techniques.

8.0.4 Traveling Salesman Problem

The *Traveling Salesman Problem* (TSP) consists in finding a Hamiltonian circuit of minimum cost on a given directed graph $G = (V, A)$. This problem arises naturally when it is necessary to distribute a given product to a set of locations, or when we need to optimally sequence a set of jobs. In some cases, the problem can be analogously defined on a undirected graph; this happens when the cost associated with an arc does not depend on its orientation.

A possible ILP model uses the following decision variables

$$
x_{ij} = \begin{cases} 1 & \text{if arc } (i,j) \in A \text{ is chosen in the optimal circuit} \\ 0 & \text{otherwise.} \end{cases}
$$

We thus obtain the following model:

$$
\min \underbrace{\sum_{(i,j)\in A} c_{ij} x_{ij}}_{\text{circuit cost}} \tag{8.2a}
$$

$$
\underbrace{\sum_{(i,j)\in\delta^-(j)} x_{ij}}_{\text{one arc entering } j} = 1 \ , \ j \in V \tag{8.2b}
$$

$$
\underbrace{\sum_{(i,j)\in\delta^+(i)} x_{ij}}_{\text{one arc leaving } i} = 1 \ , \ i \in V \tag{8.2c}
$$

$$
\underbrace{\sum_{(i,j)\in\delta^+(S)} x_{ij}}_{\text{reachability from 1}} \geq 1 \ , \ S \subset V : 1 \in S \tag{8.2d}
$$

$$
x_{ij} \geq 0 \ \text{integer} \ , \ (i,j) \in A. \tag{8.2e}
$$

Note that condition $x_{ij} \leq 1$ is implied by (8.2b) and (8.2e). Every solution that satisfies equations (8.2b)-(8.2c) corresponds to a family C of disjoint circuits covering once and only once all vertices of G. Constraints (8.2d) impose the connectivity of the solution, and require that every vertex $t \neq 1$ is reachable from vertex 1. In this way, we exclude the circuits that do not pass through vertex 1, thus imposing that family C contains a single (Hamiltonian) circuit; see Figure 8.2.

Model (8.2a)-(8.2e) can be solved using the *branch-and-cut* technique. In this case, it is necessary to initially remove the $O(2^n)$ connectivity constraints (8.2d), and then iteratively add some of them, chosen among those that are (maximally) violated by solution $\mathbf{x}^*$ of the current continuous relaxation. To that end, we need to solve the following

Separation Problem for constraints (8.2d) *Given a point* $\mathbf{x}^* \geq 0$, *identify a set* $S^* \subset V$ *such that* $1 \in S^*$ *and* $\gamma^* := \sum_{(i,j)\in\delta^+(S^*)} x_{ij}^*$ *is minimum. If* $\gamma^* \geq 1$, *then no violated constraints (8.2d) exist; otherwise,* S^* *corresponds to a constraint (8.2d) most violated by* $\mathbf{x}^*$.

If we interpret x_{ij}^* as a capacity associated with each arc $(i,j) \in A$, then set S^* corresponds to a cut $(S^*, V \setminus S^*)$ of minimum capacity in network $G^* = (V, A^* := \{(i,j) \in A : x_{ij}^* > 0\})$ with source vertex $s := 1$. Even if in this case the sink vertex t is not defined a priori, it is possible to determine cut $(S^*, V \setminus S^*)$ by selecting each vertex

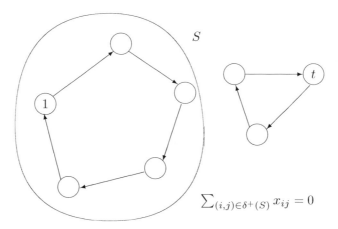

Figure 8.2: Solution without constraints (8.2d): vertex t is not reachable from vertex 1

$t \in V \setminus \{1\}$, in turn as sink, and identifying through a MAX-FLOW algorithm a cut of minimum capacity $(S_t^*, V \setminus S_t^*)$ among those with $1 \in S_t^*$ and $t \notin S_t^*$. This procedure has the advantage of generating a number of violated constraints (8.2d) to add to the current relaxation—potentially, a constraint for each $t \in V \setminus \{1\}$.

8.0.5 Steiner Tree Problem

Let $G = (V, A)$ be a directed and arc-weighted graph, let r be a given *root* vertex, and let $T \subseteq V \setminus \{r\}$ be a subset of *terminal* nodes. The *Steiner Tree Problem* consists in identifying a partial arborescence of root r reaching at minimum cost all vertices of T, possibly using one or more vertices in $V \setminus (T \cup \{r\})$; see Figure 8.3.

This model can be applied, for example, to telecommunications networks in which a signal is to be sent to certain receivers, possibly passing through intermediate stations. In some situations, it is possible that the costs associated with the arcs are negative ("prize" to reach a station).

Using variables

$$x_{ij} = \begin{cases} 1 & \text{if arc from } (i,j) \in A \text{ is selected} \\ 0 & \text{otherwise} \end{cases}$$

we obtain the following ILP model:

$$\min \underbrace{\sum_{(i,j) \in A} c_{ij} x_{ij}}_{\text{solution cost}} \tag{8.3a}$$

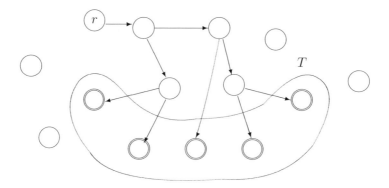

Figure 8.3: Steiner Tree Problem

$$
\underbrace{\sum_{(i,j)\in\delta^-(j)} x_{ij}}_{\text{n. of arcs entering } j}
\begin{cases}
= 1 & \text{for all } j \in T \\
= 0 & \text{for } j = r \\
\leq 1 & \text{for all } j \in V \setminus (T \cup \{r\})
\end{cases}
\tag{8.3b}
$$

$$\sum_{(i,j)\in\delta^+(S)} x_{ij} \geq \sum_{(i,t)\in\delta^-(t)} x_{it} \text{ , for all } S \subset V : r \in S, \text{ and for all } t \in V \setminus S \tag{8.3c}$$

$$x_{ij} \geq 0 \text{ integer , } (i,j) \in A. \tag{8.3d}$$

Constraints (8.3b) impose that root r does not have entering arcs, that the terminal nodes are reached by a single arc, and that the remaining nodes $j \notin T \cup \{r\}$ are isolated ($\sum_{(i,j)\in\delta^-(j)} x_{ij} = 0$) or are reached by a single arc ($\sum_{(i,j)\in\delta^-(j)} x_{ij} = 1$). Constraints (8.3c) impose that all non-isolated nodes t (those with $\sum_{(i,t)\in\delta^-(t)} x_{it} = 1$) are effectively reachable from the root: each cut $\delta^+(S)$ with $r \in S$ and $t \notin S$ must be crossed by, at least, one chosen arc.

As in the TSP case, model (8.3a)-(8.3d) can be solved by means of the *branch-and-cut* technique using a separation scheme for constraints (8.3c) based on the computation, for each set $t \in V \setminus \{r\}$, of a cut with minimum capacity $(S_t^*, V \setminus S_t^*)$ with $r \in S_t^*$ and $t \notin S_t^*$ on the network with capacity x_{ij}^* on the arcs: if the capacity of this cut is lower than $\sum_{(i,t)\in\delta^-(t)} x_{it}^*$, then the corresponding constraint (8.3c) is violated.

8.0.6   Plant Location Problem

Let n be a set of *users* of a certain service, and let m be the possible *locations* identified to activate the service. Each location has a *fixed activation cost* d_j. Each user $i \in \{1, \ldots, n\}$ has also to pay a cost c_{ij} to reach the nearest activated location j. The *Plant Location Problem* consists in identifying the locations in which to activate the service, so as to minimize the sum of the fixed and connection costs. This problem can be represented by an undirected bipartite graph $G = (V, E)$ in which a set of vertices corresponds to the users and the other set to the possible locations (see Figure 8.4).

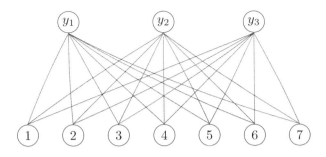

Figure 8.4: Plant Location Problem (3 locations, 7 users)

Introducing a variable for each edge $[i, j] \in E$

$$x_{ij} = \begin{cases} 1 & \text{if user } i \text{ connects to location } j \\ 0 & \text{otherwise} \end{cases}$$

and a variable for each location $j \in \{1, \ldots, m\}$

$$y_j = \begin{cases} 1 & \text{if location } j \text{ is activated} \\ 0 & \text{otherwise,} \end{cases}$$

we obtain the following ILP model:

$$\min \underbrace{\sum_{j=1}^{m} d_j y_j}_{\text{fixed cost}} + \underbrace{\sum_{i=1}^{n} \sum_{j=1}^{m} c_{ij} x_{ij}}_{\text{connection cost}} \tag{8.4a}$$

$$\underbrace{\sum_{j=1}^{m} x_{ij} = 1}_{\text{each user connects to exactly one location}}, \qquad i \in \{1, \ldots, n\} \tag{8.4b}$$

$$\underbrace{x_{ij} \leq y_j}_{\text{connection } i - j \ \Rightarrow \ \text{loc. } j \text{ activated}}, \qquad i \in \{1, \ldots, n\} \ , \ j \in \{1, \ldots, m\} \tag{8.4c}$$

$$x_{ij} \geq 0 \text{ integer} \ , \ i \in \{1, \ldots, n\} \ , \ j \in \{1, \ldots, m\} \tag{8.4d}$$

$$0 \leq y_j \leq 1 \text{ integer} \ , \ j \in \{1, \ldots, m\}. \tag{8.4e}$$

where constraints (8.4c) express the logical consistency of variables x_{ij} and y_j.

A possible solution technique is *branch-and-bound*, or *branch-and-cut* if constraints (8.4c) are initially eliminated and hence introduced, if necessary, through a simple separation algorithm; see the Index Selection Problem on the related section of Page 115.

8.0.7   Set Covering/Partitioning Problem

Given a $m \times n$ matrix A with components $a_{ij} \in \{0,1\}$ and a vector $\mathbf{c} \in \Re^n$, the *Set Covering Problem* (SCP) is defined as:

$$\begin{cases} \min \mathbf{c}^T \mathbf{x} \\ A\mathbf{x} \geq \mathbf{1} \\ 0 \leq \mathbf{x} \leq \mathbf{1} \text{ integer} \end{cases}$$

where $\mathbf{1}$ is a vector with all components equal to 1. By replacing in the model "$\geq$" with "$=$" we obtain the *Set Partitioning Problem* (SPP).

An interpretation of the SCP model is the following. Let a ground set of m items $(1, \ldots, m)$, associated with the rows of matrix A, and n subsets of the ground set be given. Each subset is defined as

$$I_j := \{i \in \{1, \ldots, m\} : a_{ij} = 1\},$$

i.e., it contains (*covers*) the items associated with the rows i with $a_{ij} = 1$, and has a given cost c_j. The SCP problem consists in finding the family of subsets I_j of minimum overall cost that covers all m items (each item has to belong to at least one selected subset).

In the SPP variant, the selected subsets I_j are also required to be pairwise disjoint (each item has to belong to one and only one subset), thus defining a partition of the ground set $\{1, \ldots, m\}$.

SCP/SPP has numerous and important applications, e.g., to define the drivers' shifts for public transport (*Crew Scheduling Problem*). In this domain, the m items correspond to m given *trips* (e.g., bus trips), each of which is defined by an origin time/location and destination time/location. A *shift* is a sequence of trips that can be assigned to a single driver, without violating any regulation/guideline. Each shift thus corresponds to a subset I_j of trips, and has an associated cost of c_j. The Crew Scheduling Problem consists then in selecting a set of shifts of minimum cost covering all m trips.

Enumerating all possible shifts (limiting, if necessary, to the most promising ones) it is possible to formulate the Crew Scheduling Problem as an SCP, defining the elements of matrix A as $a_{ij} := 1$ if the j-th shift covers the i-th trip, and $a_{ij} := 0$ otherwise. Note that a solution could select two shifts that cover the same trip i, in which case one of the two drivers will travel as passenger during trip i. If this is deemed unacceptable, one has to resort to the more constrained (and more challenging) SPP.

The solution of SCP/SPP can be tackled by *branch-and-bound*. In some applications, however, the matrix A is so large ($m \approx 2000 - 5000$, $n \approx 100.000 - 1.000.000$) that the model cannot be solved to proven optimality in a reasonable amount of time. In these cases one can to resort to (even quite sophisticated) heuristic algorithms, that can find in reasonable time feasible solutions close to the optimal one, albeit without guarantees.

Chapter 9

Exercises

9.1 Linear Programming Models

◇ **Exercise 9–1:** Write a linear programming model for the following problem. A company wants to invest no more than 40 millions in 2015, and no more than 20 millions in 2016. Five possible investments have been identified, each of which can be made in fraction from 0 to 100%; for example, deciding to make investment A at 10 %, the company will pay 1.1 millions in 2015 and 0.3 millions in 2016, and the net gain will be equal to 1.3 millions at the end of 2016. The aim is to maximize the total net gain by the end of 2016.

INVESTMENT	A	B	C	D	E
Share in 2015 (millions)	11	53	5	5	29
Share in 2016 (millions)	3	6	5	1	34
Net gain end 2016	13	16	16	14	39

◇ **Exercise 9–2:** A refinery has 10 million barrels of crude oil A and 6 million barrels of crude oil B. The refinery has 3 plants to produce gasoline (profit of 2 € /barrel) and naphtha (profit 1 € /barrel) with the performance characteristics shown in the following figure.

plant	input A	input B	output gasoline	output naphtha
1	3	5	4	3
2	1	1	1	1
3	5	3	3	4

For example, if 3 barrels of crude oil A and 5 barrels of crude oil B enter plant 1, it is possible to get 4 barrels of gasoline and 3 barrels of naphtha. The aim is to maximize the total profit.

⋄ **Exercise 9–3:** A confectionery factory produces milk chocolate and dark chocolate. The sales profits, the quantities of basic ingredients (milk and cocoa) for the production of dark and milk chocolate, and the required processing hours are shown in the following table. The data refer to the production of one ton of chocolate.

type	profit	cocoa needed	milk needed	time (h)
milk chocolate	3 ML	0.1	0.6	8
dark chocolate	2 ML	0.1	0.5	3

The factory has 6 tons of milk, 1.1 tons of cocoa, and the machines are available for no more than 48 hours in total. The aim is to maximize the overall profit.

9.2   Linear Programming

⋄ **Exercise 9–4:** Solve with the primal simplex algorithm (Bland's rule) the following Linear Programming problem. Use the two-phase method, adding two artificial variables during phase 1.

$$\begin{cases} \min & x_1 & -2x_2 \\ & 2x_1 & & +3x_3 & = & 1 \\ & 3x_1 & +2x_2 & -x_3 & = & 5 \\ & x_1, & x_2, & x_3 & \geq & 0 \end{cases}$$

⋄ **Exercise 9–5:** Solve the following Linear Programming problem by means of the primal simplex algorithm (two-phase method, Bland's rule).

$$\begin{cases} \min & x_1 & +x_2 & +2x_3 & +4x_4 \\ & & 2x_2 & & -3x_4 & = & 1 \\ & x_1 & & & -x_4 & = & 0 \\ & -x_1 & & +x_3 & & = & 1 \\ & x_1, & x_2, & x_3, & x_4 & \geq & 0 \end{cases}$$

⋄ **Exercise 9–6:** Solve the following problem both graphically and by means of the primal simplex algorithm (Bland's rule, without phase 1):

$$
\begin{cases}
\max & 3x_1 & +5x_2 \\
& 7x_1 & -3x_2 & \leq & 2 \\
& -x_1 & +x_2 & \leq & 1 \\
& x_1 & -3x_2 & \leq & 3 \\
& x_1, & x_2 & \geq & 0
\end{cases}
$$

⋄ **Exercise 9–7:** Solve with the simplex algorithm (two-phase method, Bland's rule) the following Linear Programming problem:

$$
\begin{cases}
\min & x_1 & -2x_2 \\
& 2x_1 & +x_2 & -3x_3 & = & 1 \\
& 3x_1 & +2x_2 & -x_3 & = & 5 \\
& x_1, & x_2, & x_3 & \geq & 0
\end{cases}
$$

9.3 Duality

⋄ **Exercise 9–8:** Write the dual problem associated with the following Linear Programming problem:

$$
\begin{cases}
\min & & 2x_2 & +x_3 & -3x_4 \\
& x_1 & -x_2 & & +2x_4 & \geq & 2 \\
& & 2x_2 & +x_3 & & = & 4 \\
& 2x_1 & & -x_3 & +x_4 & \leq & 1 \\
& x_1 & & & & \geq & 0 \\
& & x_2 & & & \geq & 0 \\
& & & x_3 & & & \text{free} \\
& & & & x_4 & & \text{free}
\end{cases}
$$

⋄ **Exercise 9–9:** Write the dual problem associated with the following Linear Programming problem:

$$
\left\{
\begin{array}{llllll}
\min & x_1 & -x_2 & & +x_4 & \\
& x_1 & +x_2 & -x_3 & & \geq 2 \\
& & x_2 & +x_3 & & \leq 1 \\
& x_1 & & & +x_4 & = 5 \\
& x_1 & & & & \geq 0 \\
& & x_2 & & & \geq 0 \\
& & & x_3 & & \text{free} \\
& & & & x_4 & \text{free}
\end{array}
\right.
$$

⋄ **Exercise 9–10:** Solve with the dual simplex algorithm the Linear Programming problem $\min\{\mathbf{c}^T\mathbf{x} : A\mathbf{x} = \mathbf{b}, \mathbf{x} \geq 0\}$ associated with the data shown below. In case of ambiguity in the choice of the pivot element, do *not* use Bland's rule, but choose the row/column of minimum *index*.

$$
\mathbf{c}^T = \begin{bmatrix} 0 & 0 & 0 & 4 & 3 & 1 & 0 \end{bmatrix}
$$

$$
A = \begin{bmatrix}
0 & 1 & 0 & -5 & 1 & 3 & 0 \\
1 & 0 & 0 & -1 & 0 & 4 & 0 \\
0 & 0 & 1 & 0 & -1 & 3 & 0 \\
0 & 0 & 0 & 0 & 2 & -3 & 1
\end{bmatrix}
\qquad
\mathbf{b} = \begin{bmatrix} -1 \\ -5 \\ -3 \\ -5 \end{bmatrix}
$$

9.4   Integer Linear Programming

⋄ **Exercise 9–11:** Prove the total unimodularity of the following matrix:

$$
\begin{bmatrix}
1 & 0 & 0 & 0 & 1 & 0 & 0 & 0 \\
0 & 1 & 0 & 0 & 0 & 0 & -1 & 0 \\
0 & 1 & -1 & 1 & 1 & 0 & 0 & 0 \\
0 & 0 & 0 & 0 & 0 & 0 & 1 & 0 \\
0 & 0 & 1 & 0 & 0 & 1 & 0 & 1 \\
0 & 0 & 0 & -1 & 0 & 0 & 0 & -1 \\
0 & 0 & 0 & 1 & 0 & 0 & 0 & 0 \\
-1 & 0 & 0 & 0 & 0 & 0 & 1 & 0
\end{bmatrix}
$$

⋄ **Exercise 9–12:** Write the Gomory's cuts (both integer and fractional) that can be obtained from tableau:

		x_1	x_2	x_3	x_4	x_5
$-z$	10	0	$\frac{2}{3}$	0	0	$\frac{1}{2}$
x_3	$\frac{1}{3}$	0	$-\frac{1}{3}$	1	0	$\frac{1}{4}$
x_1	3	1	$\frac{1}{5}$	0	0	$-\frac{1}{3}$
x_4	$\frac{7}{5}$	0	$\frac{4}{3}$	0	1	$-\frac{4}{3}$

⋄ **Exercise 9–13:** Solve the following Integer Linear Programming problem through Gomory's cutting plane algorithm (always choose the generating row, different from 0, of minimum index).

$$\left\{ \begin{array}{llll} \max & 3x_1 & +2x_2 & \\ & 2x_1 & +3x_2 & \leq & 3 \\ & 3x_1 & +2x_2 & \leq & 4 \\ & x_1, & x_2 & \geq & 0 \text{ integer} \end{array} \right.$$

⋄ **Exercise 9–14:** Solve the following ILP problem by means of Gomory's cutting plane algorithm (always choose the generating row, different from 0, of minimum index).

$$\left\{ \begin{array}{llll} \max & 4x_1 & +5x_2 & \\ & 2x_1 & +2x_2 & \leq & 8 \\ & x_1 & +3x_2 & \leq & 7 \\ & 2x_1 & +x_2 & \leq & 5 \\ & x_1, & x_2 & \geq & 0 \text{ integer} \end{array} \right.$$

⋄ **Exercise 9–15:** Solve the following ILP problem by means of Gomory's cutting plane algorithm (always choose the generating row, different from 0, of minimum index).

$$\left\{ \begin{array}{llll} \max & 3x_1 & +2x_2 & \\ & 2x_1 & +x_2 & \leq & 7 \\ & 3x_1 & +2x_2 & \leq & 8 \\ & x_1 & +x_2 & \leq & 6 \\ & x_1, & x_2 & \geq & 0 \text{ integer} \end{array} \right.$$

⋄ **Exercise 9–16:** Solve the following ILP problem by means of Gomory's cutting plane algorithm (always choose the generating row, different from 0, of minimum index).

$$\left\{ \begin{array}{llll} \max & 2x_1 & +x_2 & \\ & 3x_1 & -2x_2 & \leq & 0 \\ & x_1 & +2x_2 & \leq & 6 \\ & x_1, & x_2 & \geq & 0 \text{ integer} \end{array} \right.$$

⋄ **Exercise 9–17:** Solve with Gomory's cuts method the following ILP problem:

$$\begin{cases} \max & x_1 & +x_2 \\ & 3x_1 & -2x_2 & \leq & 6 \\ & & x_2 & \leq & 2 \\ & x_1, & x_2 & \geq & 0 \text{ interi} \end{cases}$$

⋄ **Exercise 9–18:** Solve with Gomory's cuts method the following ILP problem:

$$\begin{cases} \max & 4x_1 & +5x_2 \\ & 3x_1 & +2x_2 & \leq & 10 \\ & x_1 & +4x_2 & \leq & 11 \\ & x_1, & x_2 & \geq & 0 \text{ interi} \end{cases}$$

⋄ **Exercise 9–19:** Solve with Gomory's cuts method the following ILP problem:

$$\begin{cases} \min & -5x_1 & -2x_2 \\ & 2x_1 & +2x_2 & \leq & 9 \\ & 3x_1 & +x_2 & \leq & 11 \\ & x_1, & x_2 & \geq & 0 \text{ interi} \end{cases}$$

⋄ **Exercise 9–20:** Solve with Gomory's cuts method the following ILP problem:

$$\begin{cases} \max & x_1 \\ & 2x_1 & +2x_2 & \leq & 1 \\ & 4x_1 & & \geq & 1 \\ & x_1, & x_2 & \geq & 0 \text{ interi} \end{cases}$$

⋄ **Exercise 9–21:** At root node of the branching tree (*branch-and-bound*) the following tableau was obtained:

		x_1	x_2	x_3	x_4	x_5	x_6	x_7	x_8
$-z$	$-\frac{25}{3}$	0	$\frac{4}{3}$	$\frac{19}{6}$	$\frac{9}{2}$	0	0	0	$\frac{7}{6}$
x_5	1	0	1	$-\frac{1}{2}$	$-\frac{3}{2}$	1	0	0	$\frac{3}{2}$
x_1	$\frac{11}{3}$	1	$-\frac{2}{3}$	$-\frac{1}{3}$	0	0	0	0	$\frac{2}{3}$
x_6	$\frac{2}{3}$	0	$\frac{1}{3}$	$\frac{1}{6}$	$-\frac{1}{2}$	0	1	0	$\frac{7}{6}$
x_7	1	0	-3	$\frac{1}{2}$	$\frac{9}{2}$	0	0	1	$-\frac{15}{2}$

Choose x_1 as branching variable, and process the first level of the branching tree (i.e., the two child nodes of the root). For each node, solve the corresponding linear programming problem by means of the dual simplex algorithm, starting from the optimal tableau of the parent node. After having processed the second node, stop processing the branching tree and report the range within which the value of the optimal solution lies.

9.5 Graph Theory

$\diamond$ **Exercise 9–22:** Compute the (spanning) tree of *maximum* cost in the graph in figure, using Kruskal's algorithm.

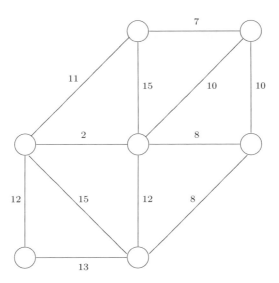

$\diamond$ **Exercise 9–23:** Compute with the Ford-Fulkerson algorithm the maximum flow from 1 to 7 and the cut of minimum capacity of the network shown in the figure. At each iteration of the inner loop, extend the labels starting from the smallest labeled yet unprocessed node. At the last iteration, highlight the cut of minimum capacity and the optimal flow associated with each arc.

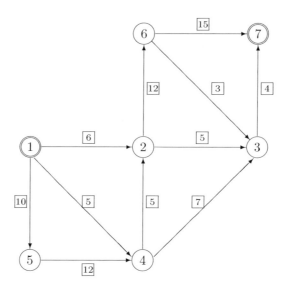

◇ **Exercise 9–24:** Compute with the Ford-Fulkerson algorithm the maximum flow from 1 to 8 and the cut of minimum capacity of the network in the figure. At each iteration of the inner loop, extend the labels starting from the smallest labeled yet unprocessed node. At the last iteration, highlight the cut of minimum capacity.

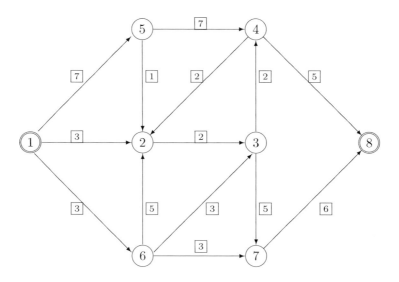

◇ **Exercise 9–25:** Write an Integer Linear Programming model for the following graph theory problem.

Let $G = (V, A)$ be a complete directed graph with costs $c_{ij} \geq 0$ on the arcs. Given two distinct vertices v_1 and v_2 and a subset $Q \subseteq V \setminus \{v_1, v_2\}$, we want to identify a simple path of minimum cost from v_1 to v_2 in which at least half of the visited vertices belongs to Q.

◇ **Exercise 9–26:** Write an Integer Linear Programming model for the following graph theory problem.

Let $G = (V, A)$ be a complete directed graph with costs $c_{ij} \geq 0$ on the arcs. Given two distinct vertices v_1 and v_2 and a subset $Q \subseteq V \setminus \{v_1, v_2\}$, we want to identify a simple path of minimum cost from v_1 to v_2 which visits at least half of the vertices belonging to Q.

◇ **Exercise 9–27:** Write an Integer Linear Programming model for the following graph theory problem.

Let $G = (V, E)$ be a given undirected graph in which the set of edges E is partitioned in two (given) sets E_1 and E_2. For each edge $e \in E$, two real values c_e (edge cost) and w_e (edge weight) are given. We want to find a spanning tree of overall minimum cost satisfying the following requirements:

(1) the overall weight does not have to be lower than a given value W;

(2) the number of chosen edges in E_1 has to be lower than the number of chosen edges in E_2.

◇ **Exercise 9–28:** Write an Integer Linear Programming model for the following graph theory problem.

Let $G = (V, E)$ be a given undirected graph in which the set of edges E is partitioned in two (given) sets E_1 and E_2. For each edge e, two real values c_e (edge cost) and w_e (edge weight) are given. We want to find a spanning tree of overall *maximum* cost satisfying the following requirements:

(1) the average weight of the edges chosen in E_1 does not have to be greater than a given value W;

(2) at least half of the chosen edges must belong to E_1.

◇ **Exercise 9–29:** Write a Linear Programming model for the following graph theory problem.

Let $G = (V, E)$ be an undirected graph with costs c_e on the edges, in which an integer given value b_v is associated with each $v \in V$. A *b-matching* is a partial graph $G' = (V, E')$ of G such that $d_{G'}(v) = b_v$ for all $v \in V$. We want to identify a *b-matching* of overall minimum cost in G, if it exists.

Chapter 10

Solutions

◇ **Exercise 9–1:** Let

$$x_i = \text{investment fraction } i \text{ to be made } (i = 1, \ldots, 5)$$

$$
\begin{cases}
\max & 13x_1 & +16x_2 & +16x_3 & +14x_4 & +39x_5 & & & \text{(gain)} \\
& 11x_1 & +53x_2 & +5x_3 & +5x_4 & +29x_5 & \leq & 40 & \text{(budget 2015)} \\
& 3x_1 & +6x_2 & +5x_3 & +x_4 & +34x_5 & \leq & 20 & \text{(budget 2016)} \\
& x_1 & & & & & \leq & 1 \\
& & x_2 & & & & \leq & 1 \\
& & & x_3 & & & \leq & 1 \\
& & & & x_4 & & \leq & 1 \\
& & & & & x_5 & \leq & 1 \\
& x_1, & x_2, & x_3, & x_4, & x_5 & \geq & 0
\end{cases}
$$

◇ **Exercise 9–2:** Let

$$x_i = \text{production level of plant } i \ (i = 1, 2, 3)$$

$$
\begin{cases}
\max & 2(4x_1 + x_2 + 3x_3) + (3x_1 + x_2 + 4x_3) = \\
& = 11x_1 & +3x_2 & +10x_3 & & \text{(profit)} \\
& 3x_1 & +x_2 & +5x_3 & \leq 10 & \text{(crude oil A)} \\
& 5x_1 & +x_2 & +3x_3 & \leq 6 & \text{(crude oil B)} \\
& x_1, & x_2, & x_3 & \geq 0
\end{cases}
$$

⋄ **Exercise 9–3:** Let

$$x_1 = \text{quantity of milk chocolate produced}$$
$$x_2 = \text{quantity of dark chocolate produced}$$

$$
\begin{cases}
\max \quad 3x_1 & +2x_2 & & & \text{(profit)} \\
6x_1 & +5x_2 & \leq & 60 & \text{(milk)} \\
x_1 & +x_2 & \leq & 11 & \text{(cocoa)} \\
8x_1 & +3x_2 & \leq & 48 & \text{(processing)} \\
x_1, & x_2 & \geq & 0 &
\end{cases}
$$

⋄ **Exercise 9–4:**

Phase 1:

		x_1	x_2	x_3	x_4	x_5
$-w$	-6	-5	-2	-2	0	0
x_4	1	②	0	3	1	0
x_5	5	3	2	-1	0	1

		x_1	x_2	x_3	x_4	x_5
$-w$	$-\frac{7}{2}$	0	-2	$\frac{11}{2}$	$\frac{5}{2}$	0
x_1	$\frac{1}{2}$	1	0	$\frac{3}{2}$	$\frac{1}{2}$	0
x_5	$\frac{7}{2}$	0	②	$-\frac{11}{2}$	$-\frac{3}{2}$	1

		x_1	x_2	x_3	x_4	x_5
$-w$	0	0	0	0	1	1
x_1	$\frac{1}{2}$	1	0	$\frac{3}{2}$	$\frac{1}{2}$	0
x_2	$\frac{7}{4}$	0	1	$-\frac{11}{4}$	$-\frac{3}{4}$	$\frac{1}{2}$

Phase 2:

		x_1	x_2	x_3
$-z$	3	0	0	-7
x_1	$\frac{1}{2}$	1	0	③ $\frac{3}{2}$
x_2	$\frac{7}{4}$	0	1	$-\frac{11}{4}$

		x_1	x_2	x_3
$-z$	$\frac{16}{3}$	$\frac{14}{3}$	0	0
x_3	$\frac{1}{3}$	$\frac{2}{3}$	0	1
x_2	$\frac{8}{3}$	$\frac{11}{6}$	1	0

◇ **Exercise 9–5:**

Phase 1:

		x_1	x_2	x_3	x_4	x_5	x_6	x_7
$-w$	-2	0	-2	-1	4	0	0	0
x_5	1	0	②	0	-3	1	0	0
x_6	0	1	0	0	-1	0	1	0
x_7	1	-1	0	1	0	0	0	1

		x_1	x_2	x_3	x_4	x_5	x_6	x_7
$-w$	-1	0	0	-1	1	1	0	0
x_2	$\frac{1}{2}$	0	1	0	$-\frac{3}{2}$	$\frac{1}{2}$	0	0
x_6	0	1	0	0	-1	0	1	0
x_7	1	-1	0	①	0	0	0	1

		x_1	x_2	x_3	x_4	x_5	x_6	x_7
$-w$	0	-1	0	0	1	1	0	1
x_2	$\frac{1}{2}$	0	1	0	$-\frac{3}{2}$	$\frac{1}{2}$	0	0
x_6	0	①	0	0	-1	0	1	0
x_3	1	-1	0	1	0	0	0	1

		x_1	x_2	x_3	x_4	x_5	x_6	x_7
$-w$	0	0	0	0	0	1	1	1
x_2	$\frac{1}{2}$	0	1	0	$-\frac{3}{2}$	$\frac{1}{2}$	0	0
x_1	0	1	0	0	-1	0	1	0
x_3	1	0	0	1	-1	0	1	1

Phase 2:

		x_1	x_2	x_3	x_4
$-z$	$-\frac{5}{2}$	0	0	0	$\frac{17}{2}$
x_2	$\frac{1}{2}$	0	1	0	$-\frac{3}{2}$
x_1	0	1	0	0	-1
x_3	1	0	0	1	-1

◇ **Exercise 9–6:**

		x_1	x_2	x_3	x_4	x_5
$-z$	0	-3	-5	0	0	0
x_3	2	⑦	-3	1	0	0
x_4	1	-1	1	0	1	0
x_5	3	1	-3	0	0	1

		x_1	x_2	x_3	x_4	x_5
$-z$	$\frac{6}{7}$	0	$-\frac{44}{7}$	$\frac{3}{7}$	0	0
x_1	$\frac{2}{7}$	1	$-\frac{3}{7}$	$\frac{1}{7}$	0	0
x_4	$\frac{9}{7}$	0	$\frac{4}{7}$	$\frac{1}{7}$	1	0
x_5	$\frac{19}{7}$	0	$-\frac{18}{7}$	$-\frac{1}{7}$	0	1

		x_1	x_2	x_3	x_4	x_5
$-z$	15	0	0	2	11	0
x_1	$\frac{5}{4}$	1	0	$\frac{1}{4}$	$\frac{3}{4}$	0
x_2	$\frac{9}{4}$	0	1	$\frac{1}{4}$	$\frac{7}{4}$	0
x_5	$\frac{17}{2}$	0	0	$\frac{1}{2}$	$\frac{9}{2}$	1

Graphical representation:

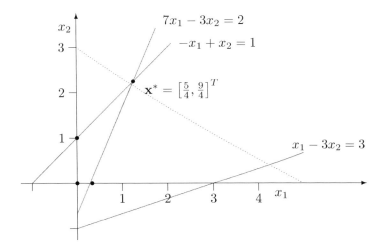

◇ **Exercise 9–7:**

Phase 1.

		x_1	x_2	x_3	x_4	x_5
$-w$	-6	-5	-3	4	0	0
x_4	1	②	1	-3	1	0
x_5	5	3	2	-1	0	1

		x_1	x_2	x_3	x_4	x_5
$-w$	$-\frac{7}{2}$	0	$-\frac{1}{2}$	$-\frac{7}{2}$	$\frac{5}{2}$	0
x_1	$\frac{1}{2}$	1	$\frac{1}{2}$	$-\frac{3}{2}$	$\frac{1}{2}$	0
x_5	$\frac{7}{2}$	0	$\frac{1}{2}$	$\frac{7}{2}$	$-\frac{3}{2}$	1

		x_1	x_2	x_3	x_4	x_5
$-w$	-3	1	0	-5	3	0
x_2	1	2	1	-3	1	0
x_5	3	-1	0	⑤	-2	1

		x_1	x_2	x_3	x_4	x_5
$-w$	0	0	0	0	1	1
x_2	$\frac{14}{5}$	$\frac{7}{5}$	1	0	$-\frac{1}{5}$	$\frac{3}{5}$
x_3	$\frac{3}{5}$	$-\frac{1}{5}$	0	1	$-\frac{2}{5}$	$\frac{1}{5}$

Phase 2.

	x_1	x_2	x_3	
$-z$	$\frac{28}{5}$	$\frac{19}{5}$	0	0
x_2	$\frac{14}{5}$	$\frac{7}{5}$	1	0
x_3	$\frac{3}{5}$	$-\frac{1}{5}$	0	1

◇ **Exercise 9–8:**

$$
\left\{
\begin{array}{lrlrcr}
\max & 2u_1 & +4u_2 & +u_3 & & \\
     & u_1 &   &   & \geq & 0 \\
     &   & u_2   &   & \text{free} & \\
     &   &   & u_3 & \leq & 0 \\
     & u_1 &   & +2u_3 & \leq & 0 \\
     & -u_1 & +2u_2 &   & \leq & 2 \\
     &   & u_2   & -u_3 & = & 1 \\
     & 2u_1 &   & +u_3 & = & -3
\end{array}
\right.
$$

◇ **Exercise 9–9:**

$$
\left\{
\begin{array}{lrlrcr}
\max & 2u_1 & +u_2 & +5u_3 & & \\
     & u_1 &   &   & \geq & 0 \\
     &   & u_2 &   & \leq & 0 \\
     &   &   & u_3   & \text{free} & \\
     & u_1 &   & +u_3 & \leq & 1 \\
     & u_1 & +u_2 &   & \leq & -1 \\
     & -u_1 & +u_2 &   & = & 0 \\
     &   &   & u_3   & = & 1
\end{array}
\right.
$$

◇ **Exercise 9–10:**

		x_1	x_2	x_3	x_4	x_5	x_6	x_7
$-z$	0	0	0	0	4	3	1	0
x_2	-1	0	1	0	$\boxed{-5}$	1	3	0
x_1	-5	1	0	0	-1	0	4	0
x_3	-3	0	0	1	0	-1	3	0
x_7	-5	0	0	0	0	2	-3	1

		x_1	x_2	x_3	x_4	x_5	x_6	x_7
$-z$	$-\frac{4}{5}$	0	$\frac{4}{5}$	0	0	$\frac{19}{5}$	$\frac{17}{5}$	0
x_4	$\frac{1}{5}$	0	$-\frac{1}{5}$	0	1	$-\frac{1}{5}$	$-\frac{3}{5}$	0
x_1	$-\frac{24}{5}$	1	$\left(-\frac{1}{5}\right)$	0	0	$-\frac{1}{5}$	$\frac{17}{5}$	0
x_3	-3	0	0	1	0	-1	3	0
x_7	-5	0	0	0	0	2	-3	1

		x_1	x_2	x_3	x_4	x_5	x_6	x_7
$-z$	-20	4	0	0	0	3	17	0
x_4	5	-1	0	0	1	0	-4	0
x_2	24	-5	1	0	0	1	-17	0
x_3	-3	0	0	1	0	$\left(-1\right)$	3	0
x_7	-5	0	0	0	0	2	-3	1

		x_1	x_2	x_3	x_4	x_5	x_6	x_7
$-z$	-29	4	0	3	0	0	26	0
x_4	5	-1	0	0	1	0	-4	0
x_2	21	-5	1	1	0	0	-14	0
x_5	3	0	0	-1	0	1	-3	0
x_7	-11	0	0	2	0	0	3	1

The problem is infeasible.

◇ **Exercise 9–11:** After having eliminated the rows and the columns with a single element different from zero, Theorem 6.2.3 can be applied.

$$
\begin{bmatrix}
1 & 0 & 0 & 0 & 1 & 0 & 0 & 0 \\
0 & 1 & 0 & 0 & 0 & 0 & -1 & 0 \\
0 & 1 & -1 & 1 & 1 & 0 & 0 & 0 \\
0 & 0 & 0 & 0 & 0 & 0 & 1 & 0 \\
0 & 0 & 1 & 0 & 0 & 1 & 0 & 1 \\
0 & 0 & 0 & -1 & 0 & 0 & 0 & -1 \\
0 & 0 & 0 & 1 & 0 & 0 & 0 & 0 \\
-1 & 0 & 0 & 0 & 0 & 0 & 1 & 0
\end{bmatrix}
\begin{matrix}
\in I_2 \\
\in I_2 \\
\in I_1 \\
\times \\
\in I_1 \\
\in I_1 \\
\times \\
\in I_2
\end{matrix}
$$

$\diamond$ **Exercise 9–12:** The cuts are:

$$\begin{cases} -x_2 + x_3 \leq 0 \\ x_2 + x_4 - 2x_5 \leq 1 \end{cases} \quad \Rightarrow \quad \begin{cases} \frac{2}{3}x_2 + \frac{1}{4}x_5 \geq \frac{1}{3} \\ \frac{1}{3}x_2 + \frac{2}{3}x_5 \geq \frac{2}{5} \end{cases}$$

$\diamond$ **Exercise 9–13:**

		x_1	x_2	x_3	x_4
$-z$	0	-3	-2	0	0
x_3	3	2	3	1	0
x_4	4	③	2	0	1

		x_1	x_2	x_3	x_4
$-z$	4	0	0	0	1
x_3	$\frac{1}{3}$	0	$\frac{5}{3}$	1	$-\frac{2}{3}$
x_1	$\frac{4}{3}$	1	$\frac{2}{3}$	0	$\frac{1}{3}$

generating row = row 1

		x_1	x_2	x_3	x_4	x_5
$-z$	4	0	0	0	1	0
x_3	$\frac{1}{3}$	0	$\frac{5}{3}$	1	$-\frac{2}{3}$	0
x_1	$\frac{4}{3}$	1	$\frac{2}{3}$	0	$\frac{1}{3}$	0
x_5	$-\frac{1}{3}$	0	$-\frac{2}{3}$	0	$-\frac{1}{3}$	1

		x_1	x_2	x_3	x_4	x_5
$-z$	4	0	0	0	1	0
x_3	$-\frac{1}{2}$	0	0	1	$-\frac{3}{2}$	$\frac{5}{2}$
x_1	1	1	0	0	0	1
x_2	$\frac{1}{2}$	0	1	0	$\frac{1}{2}$	$-\frac{3}{2}$

		x_1	x_2	x_3	x_4	x_5
$-z$	$\frac{11}{3}$	0	0	$\frac{2}{3}$	0	$\frac{5}{3}$
x_4	$\frac{1}{3}$	0	0	$-\frac{2}{3}$	1	$-\frac{5}{3}$
x_1	1	1	0	0	0	1
x_2	$\frac{1}{3}$	0	1	$\frac{1}{3}$	0	$-\frac{2}{3}$

generating row = row 1

		x_1	x_2	x_3	x_4	x_5	
$-z$	$\frac{11}{3}$	0	0	$\frac{2}{3}$	0	$\frac{5}{3}$	0
x_4	$\frac{1}{3}$	0	0	$-\frac{2}{3}$	1	$-\frac{5}{3}$	0
x_1	1	1	0	0	0	1	0
x_2	$\frac{1}{3}$	0	1	$\frac{1}{3}$	0	$-\frac{2}{3}$	0
x_6	$-\frac{1}{3}$	0	0	$\left(-\frac{1}{3}\right)$	0	$-\frac{1}{3}$	1

		x_1	x_2	x_3	x_4	x_5	x_6
$-z$	3	0	0	0	0	1	2
x_4	1	0	0	0	1	-1	-2
x_1	1	1	0	0	0	1	0
x_2	0	0	1	0	0	-1	1
x_3	1	0	0	1	0	1	-3

◇ **Exercise 9–14:**

		x_1	x_2	x_3	x_4	x_5
$-z$	0	-4	-5	0	0	0
x_3	8	2	2	1	0	0
x_4	7	1	3	0	1	0
x_5	5	②	1	0	0	1

		x_1	x_2	x_3	x_4	x_5
$-z$	10	0	-3	0	0	2
x_3	3	0	1	1	0	-1
x_4	$\frac{9}{2}$	0	$\left(\frac{5}{2}\right)$	0	1	$-\frac{1}{2}$
x_1	$\frac{5}{2}$	1	$\frac{1}{2}$	0	0	$\frac{1}{2}$

		x_1	x_2	x_3	x_4	x_5
$-z$	$\frac{77}{5}$	0	0	0	$\frac{6}{5}$	$\frac{7}{5}$
x_3	$\frac{6}{5}$	0	0	1	$-\frac{2}{5}$	$-\frac{4}{5}$
x_2	$\frac{9}{5}$	0	1	0	$\frac{2}{5}$	$-\frac{1}{5}$
x_1	$\frac{8}{5}$	1	0	0	$-\frac{1}{5}$	$\frac{3}{5}$

generating row = row 1

		x_1	x_2	x_3	x_4	x_5	x_6
$-z$	$\frac{77}{5}$	0	0	0	$\frac{6}{5}$	$\frac{7}{5}$	0
x_3	$\frac{6}{5}$	0	0	1	$-\frac{2}{5}$	$-\frac{4}{5}$	0
x_2	$\frac{9}{5}$	0	1	0	$\frac{2}{5}$	$-\frac{1}{5}$	0
x_1	$\frac{8}{5}$	1	0	0	$-\frac{1}{5}$	$\frac{3}{5}$	0
x_6	$-\frac{1}{5}$	0	0	0	$\left(-\frac{3}{5}\right)$	$-\frac{1}{5}$	1

		x_1	x_2	x_3	x_4	x_5	x_6
$-z$	15	0	0	0	0	1	2
x_3	$\frac{4}{3}$	0	0	1	0	$-\frac{2}{3}$	$-\frac{2}{3}$
x_2	$\frac{5}{3}$	0	1	0	0	$-\frac{1}{3}$	$\frac{2}{3}$
x_1	$\frac{5}{3}$	1	0	0	0	$\frac{2}{3}$	$-\frac{1}{3}$
x_4	$\frac{1}{3}$	0	0	0	1	$\frac{1}{3}$	$-\frac{5}{3}$

generating row = row 1

		x_1	x_2	x_3	x_4	x_5	x_6	x_7
$-z$	15	0	0	0	0	1	2	0
x_3	$\frac{4}{3}$	0	0	1	0	$-\frac{2}{3}$	$-\frac{2}{3}$	0
x_2	$\frac{5}{3}$	0	1	0	0	$-\frac{1}{3}$	$\frac{2}{3}$	0
x_1	$\frac{5}{3}$	1	0	0	0	$\frac{2}{3}$	$-\frac{1}{3}$	0
x_4	$\frac{1}{3}$	0	0	0	1	$\frac{1}{3}$	$-\frac{5}{3}$	0
x_7	$-\frac{1}{3}$	0	0	0	0	$\left(-\frac{1}{3}\right)$	$-\frac{1}{3}$	1

		x_1	x_2	x_3	x_4	x_5	x_6	x_7
$-z$	14	0	0	0	0	0	1	3
x_3	2	0	0	1	0	0	0	-2
x_2	2	0	1	0	0	0	1	-1
x_1	1	1	0	0	0	0	-1	2
x_4	0	0	0	0	1	0	-2	1
x_5	1	0	0	0	0	1	1	-3

◇ **Exercise 9–15:**

		x_1	x_2	x_3	x_4	x_5
$-z$	0	-3	-2	0	0	0
x_3	7	2	1	1	0	0
x_4	8	③	2	0	1	0
x_5	6	1	1	0	0	1

		x_1	x_2	x_3	x_4	x_5
$-z$	8	0	0	0	1	0
x_3	$\frac{5}{3}$	0	$-\frac{1}{3}$	1	$-\frac{2}{3}$	0
x_1	$\frac{8}{3}$	1	$\frac{2}{3}$	0	$\frac{1}{3}$	0
x_5	$\frac{10}{3}$	0	$\frac{1}{3}$	0	$-\frac{1}{3}$	1

generating row = row 1

		x_1	x_2	x_3	x_4	x_5	x_6
$-z$	8	0	0	0	1	0	0
x_3	$\frac{5}{3}$	0	$-\frac{1}{3}$	1	$-\frac{2}{3}$	0	0
x_1	$\frac{8}{3}$	1	$\frac{2}{3}$	0	$\frac{1}{3}$	0	0
x_5	$\frac{10}{3}$	0	$\frac{1}{3}$	0	$-\frac{1}{3}$	1	0
x_6	$-\frac{2}{3}$	0	$-\frac{2}{3}$	0	$-\frac{1}{3}$	0	1

		x_1	x_2	x_3	x_4	x_5	x_6
$-z$	8	0	0	0	1	0	0
x_3	2	0	0	1	$-\frac{1}{2}$	0	$-\frac{1}{2}$
x_1	2	1	0	0	0	0	1
x_5	3	0	0	0	$-\frac{1}{2}$	1	$\frac{1}{2}$
x_2	1	0	1	0	$\frac{1}{2}$	0	$-\frac{3}{2}$

◇ **Exercise 9–16:**

		x_1	x_2	x_3	x_4
$-z$	0	-2	-1	0	0
x_3	0	③	-2	1	0
x_4	6	1	2	0	1

		x_1	x_2	x_3	x_4
$-z$	0	0	$-\frac{7}{3}$	$\frac{2}{3}$	0
x_1	0	1	$-\frac{2}{3}$	$\frac{1}{3}$	0
x_4	6	0	$\frac{8}{3}$	$-\frac{1}{3}$	1

		x_1	x_2	x_3	x_4
$-z$	$\frac{21}{4}$	0	0	$\frac{3}{8}$	$\frac{7}{8}$
x_1	$\frac{3}{2}$	1	0	$\frac{1}{4}$	$\frac{1}{4}$
x_2	$\frac{9}{4}$	0	1	$-\frac{1}{8}$	$\frac{3}{8}$

generating row = row 1

		x_1	x_2	x_3	x_4	x_5
$-z$	$\frac{21}{4}$	0	0	$\frac{3}{8}$	$\frac{7}{8}$	0
x_1	$\frac{3}{2}$	1	0	$\frac{1}{4}$	$\frac{1}{4}$	0
x_2	$\frac{9}{4}$	0	1	$-\frac{1}{8}$	$\frac{3}{8}$	0
x_5	$-\frac{1}{2}$	0	0	$-\frac{1}{4}$	$-\frac{1}{4}$	1

		x_1	x_2	x_3	x_4	x_5
$-z$	$\frac{9}{2}$	0	0	0	$\frac{1}{2}$	$\frac{3}{2}$
x_1	1	1	0	0	0	1
x_2	$\frac{5}{2}$	0	1	0	$\frac{1}{2}$	$-\frac{1}{2}$
x_3	2	0	0	1	1	-4

generating row = row 2

		x_1	x_2	x_3	x_4	x_5	x_6
$-z$	$\frac{9}{2}$	0	0	0	$\frac{1}{2}$	$\frac{3}{2}$	0
x_1	1	1	0	0	0	1	0
x_2	$\frac{5}{2}$	0	1	0	$\frac{1}{2}$	$-\frac{1}{2}$	0
x_3	2	0	0	1	1	-4	0
x_6	$-\frac{1}{2}$	0	0	0	$\left(-\frac{1}{2}\right)$	$-\frac{1}{2}$	1

		x_1	x_2	x_3	x_4	x_5	x_6
$-z$	4	0	0	0	0	1	1
x_1	1	1	0	0	0	1	0
x_2	2	0	1	0	0	-1	1
x_3	1	0	0	1	0	-5	2
x_4	1	0	0	0	1	1	-2

⋄ **Exercise 9–17:** The optimal solution is: $\mathbf{x}^* = [3, 2]^T$, $z_{PLI} = 5$

⋄ **Exercise 9–18:** The optimal solution is: $\mathbf{x}^* = [2, 2]^T$, $z_{PLI} = 18$

⋄ **Exercise 9–19:** The optimal solution is: $\mathbf{x}^* = [3, 1]^T$, $z_{PLI} = -17$

⋄ **Exercise 9–20:** No integer feasible solution exists.

⋄ **Exercise 9–21:** The optimal tableau at root node is:

		x_1	x_2	x_3	x_4	x_5	x_6	x_7	x_8
$-z$	$-\frac{25}{3}$	0	$\frac{4}{3}$	$\frac{19}{6}$	$\frac{9}{2}$	0	0	0	$\frac{7}{6}$
x_5	1	0	1	$-\frac{1}{2}$	$-\frac{3}{2}$	1	0	0	$\frac{3}{2}$
x_1	$\frac{11}{3}$	1	$-\frac{2}{3}$	$-\frac{1}{3}$	0	0	0	0	$\frac{2}{3}$
x_6	$\frac{2}{3}$	0	$\frac{1}{3}$	$\frac{1}{6}$	$-\frac{1}{2}$	0	1	0	$\frac{7}{6}$
x_7	1	0	-3	$\frac{1}{2}$	$\frac{9}{2}$	0	0	1	$-\frac{15}{2}$

Branching on $x_1 \leq 3$:

		x_1	x_2	x_3	x_4	x_5	x_6	x_7	x_8	x_9
$-z$	$-\frac{25}{3}$	0	$\frac{4}{3}$	$\frac{19}{6}$	$\frac{9}{2}$	0	0	0	$\frac{7}{6}$	0
x_5	1	0	1	$-\frac{1}{2}$	$-\frac{3}{2}$	1	0	0	$\frac{3}{2}$	0
x_1	$\frac{11}{3}$	1	$-\frac{2}{3}$	$-\frac{1}{3}$	0	0	0	0	$\frac{2}{3}$	0
x_6	$\frac{2}{3}$	0	$\frac{1}{3}$	$\frac{1}{6}$	$-\frac{1}{2}$	0	1	0	$\frac{7}{6}$	0
x_7	1	0	-3	$\frac{1}{2}$	$\frac{9}{2}$	0	0	1	$-\frac{15}{2}$	0
x_9	$-\frac{2}{3}$	0	$\frac{2}{3}$	$\frac{1}{3}$	0	0	0	0	$\left(-\frac{2}{3}\right)$	1

		x_1	x_2	x_3	x_4	x_5	x_6	x_7	x_8	x_9
$-z$	$-\frac{19}{2}$	0	$\frac{5}{2}$	$\frac{15}{4}$	$\frac{9}{2}$	0	0	0	0	$\frac{7}{4}$
x_5	$-\frac{1}{2}$	0	$\frac{5}{2}$	$\frac{1}{4}$	$\left(-\frac{3}{2}\right)$	1	0	0	0	$\frac{9}{4}$
x_1	3	1	0	0	0	0	0	0	0	1
x_6	$-\frac{1}{2}$	0	$\frac{3}{2}$	$\frac{3}{4}$	$-\frac{1}{2}$	0	1	0	0	$\frac{7}{4}$
x_7	$\frac{17}{2}$	0	$-\frac{21}{2}$	$-\frac{13}{4}$	$\frac{9}{2}$	0	0	1	0	$-\frac{45}{4}$
x_8	1	0	-1	$-\frac{1}{2}$	0	0	0	0	1	$-\frac{3}{2}$

		x_1	x_2	x_3	x_4	x_5	x_6	x_7	x_8	x_9
$-z$	-11	0	10	$\frac{9}{2}$	0	3	0	0	0	$\frac{17}{2}$
x_4	$\frac{1}{3}$	0	$-\frac{5}{3}$	$-\frac{1}{6}$	1	$-\frac{2}{3}$	0	0	0	$-\frac{3}{2}$
x_1	3	1	0	0	0	0	0	0	0	1
x_6	$-\frac{1}{3}$	0	$\frac{2}{3}$	$\frac{2}{3}$	0	$\left(-\frac{1}{3}\right)$	1	0	0	1
x_7	7	0	-3	$-\frac{5}{2}$	0	3	0	1	0	$-\frac{9}{2}$
x_8	1	0	-1	$-\frac{1}{2}$	0	0	0	0	1	$-\frac{3}{2}$

		x_1	x_2	x_3	x_4	x_5	x_6	x_7	x_8	x_9
$-z$	-14	0	16	$\frac{21}{2}$	0	0	9	0	0	$\frac{35}{2}$
x_4	1	0	-3	$-\frac{3}{2}$	1	0	-2	0	0	$-\frac{7}{2}$
x_1	3	1	0	0	0	0	0	0	0	1
x_5	1	0	-2	-2	0	1	-3	0	0	-3
x_7	4	0	3	$\frac{7}{2}$	0	0	9	1	0	$\frac{9}{2}$
x_8	1	0	-1	$-\frac{1}{2}$	0	0	0	0	1	$-\frac{3}{2}$

Branching on $x_1 \geq 4$:

		x_1	x_2	x_3	x_4	x_5	x_6	x_7	x_8	x_9
$-z$		$-\frac{25}{3}$	0	$\frac{4}{3}$	$\frac{19}{6}$	$\frac{9}{2}$	0	0	0	$\frac{7}{6}$	0
x_5		1	0	1	$-\frac{1}{2}$	$-\frac{3}{2}$	1	0	0	$\frac{3}{2}$	0
x_1		$\frac{11}{3}$	1	$-\frac{2}{3}$	$-\frac{1}{3}$	0	0	0	0	$\frac{2}{3}$	0
x_6		$\frac{2}{3}$	0	$\frac{1}{3}$	$\frac{1}{6}$	$-\frac{1}{2}$	0	1	0	$\frac{7}{6}$	0
x_7		1	0	-3	$\frac{1}{2}$	$\frac{9}{2}$	0	0	1	$-\frac{15}{2}$	0
x_9		$-\frac{1}{3}$	0	$\left(-\frac{2}{3}\right)$	$-\frac{1}{3}$	0	0	0	0	$\frac{2}{3}$	1

		x_1	x_2	x_3	x_4	x_5	x_6	x_7	x_8	x_9
$-z$		-9	0	0	$\frac{5}{2}$	$\frac{9}{2}$	0	0	0	$\frac{5}{2}$	2
x_5		$\frac{1}{2}$	0	0	-1	$-\frac{3}{2}$	1	0	0	$\frac{5}{2}$	$\frac{3}{2}$
x_1		4	1	0	0	0	0	0	0	0	-1
x_6		$\frac{1}{2}$	0	0	0	$-\frac{1}{2}$	0	1	0	$\frac{3}{2}$	$\frac{1}{2}$
x_7		$\frac{5}{2}$	0	0	2	$\frac{9}{2}$	0	0	1	$-\frac{21}{2}$	$-\frac{9}{2}$
x_2		$\frac{1}{2}$	0	1	$\frac{1}{2}$	0	0	0	0	-1	$-\frac{3}{2}$

The cost of the optimal integer solution is within the range $[9, 14]$. The branching tree is:

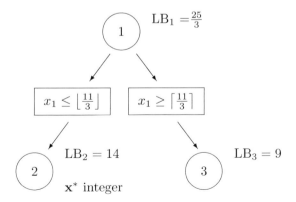

$\diamond$ **Exercise 9–22:** The maximum tree (in bold) has cost 75.

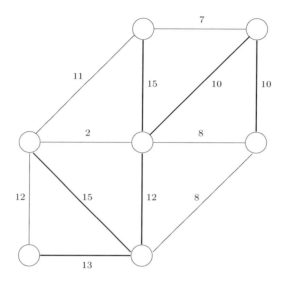

◇ **Exercise 9–23:**

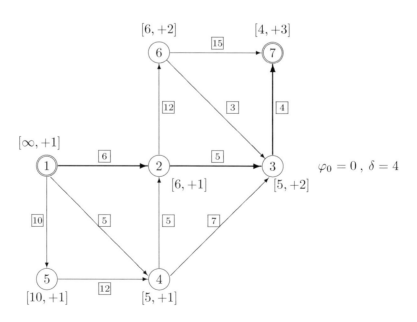

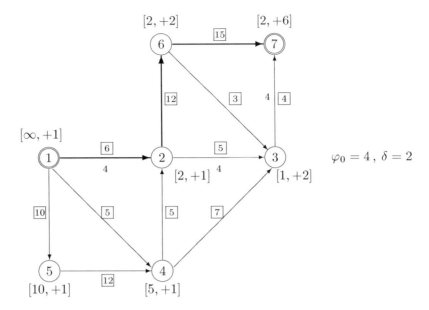

$\varphi_0 = 4 \, , \, \delta = 2$

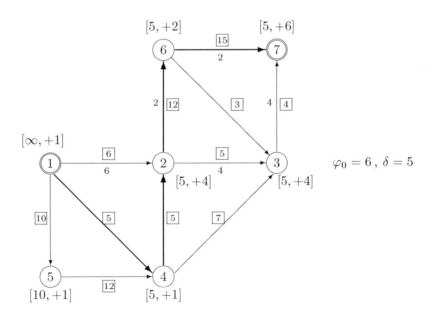

$\varphi_0 = 6 \, , \, \delta = 5$

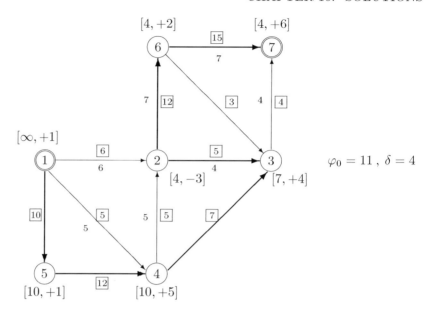

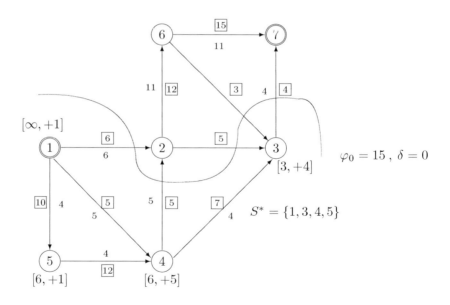

◇ **Exercise 9–24:**

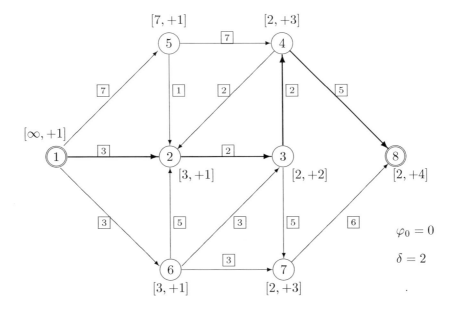

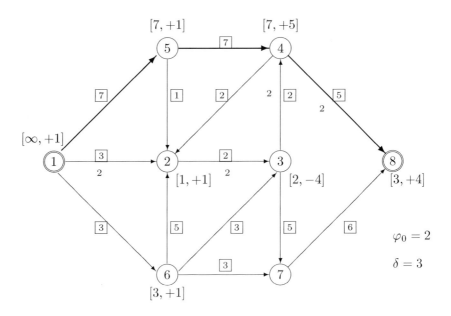

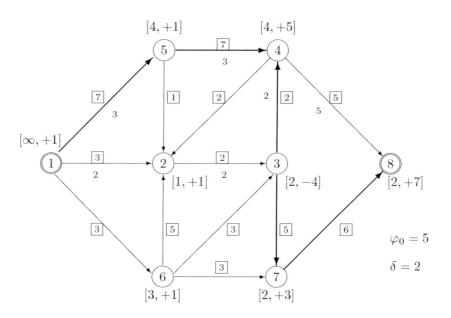

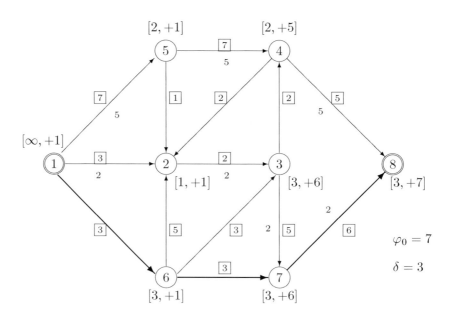

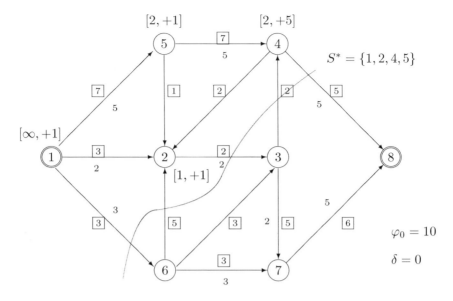

◇ **Exercise 9–25:**

$$\left\{ \begin{array}{l} \min \displaystyle\sum_{i \in V} \sum_{j \in V} c_{ij} x_{ij} \\[2ex] \displaystyle\sum_{j \in V} x_{hj} - \sum_{i \in V} x_{ih} = \left\{ \begin{array}{ll} 1 & \text{if } h = v_1 \\[1ex] -1 & \text{if } h = v_2 \\[1ex] 0 & \forall h \in V \setminus \{v_1, v_2\} \end{array} \right. \\[5ex] \displaystyle\sum_{i \in S} \sum_{j \in S} x_{ij} \leq |S| - 1 \,, \, \forall\, S \subseteq V \,, \, S \neq \emptyset \\[3ex] \displaystyle\sum_{h \in Q} \underbrace{\sum_{i \in V} x_{ih}}_{=1 \text{ if h is visited}} \geq \frac{1}{2} \left(\underbrace{\sum_{i \in V} \sum_{j \in V} x_{ij}}_{\text{n. of arcs in the path}} + 1 \right) \\[4ex] x_{ij} \geq 0 \text{ integer} \,, \, \forall\, i, j \in V \end{array} \right.$$

◇ **Exercise 9–26:**

$$
\begin{cases}
\min \sum_{i \in V} \sum_{j \in V} c_{ij} x_{ij} \\[2mm]
\sum_{j \in V} x_{hj} - \sum_{i \in V} x_{ih} = \begin{cases} 1 & \text{if } h = v_2 \\ -1 & \text{if } h = v_1 \\ 0 & \forall h \in V \setminus \{v_1, v_2\} \end{cases} \\[2mm]
\sum_{i \in S} \sum_{j \in S} x_{ij} \le |S| - 1 \,, \, \forall \, S \subseteq V \,, \, S \ne \emptyset \\[2mm]
\sum_{h \in Q} \underbrace{\sum_{i \in V} x_{ih}}_{=1 \text{ if } h \text{ is visited}} \ge \left\lceil \frac{|Q|}{2} \right\rceil \\[2mm]
x_{ij} \ge 0 \text{ integer} \,, \, \forall \, i, j \in V
\end{cases}
$$

⋄ **Exercise 9–27:**

$$
\begin{cases}
\min \sum_{e \in E} c_e x_e \\[2mm]
\sum_{e \in E} x_e = |V| - 1 \\[2mm]
\sum_{e \in E(S)} x_e \le |S| - 1 \,, \, \forall \, S \subset V \,, \, |S| \ge 3 \\[2mm]
\underbrace{\sum_{e \in E} w_e x_e}_{\text{weight of chosen edges}} \ge W \\[2mm]
\underbrace{\sum_{e \in E_1} x_e}_{\text{n. of edges in } E_1} \le \underbrace{\sum_{e \in E_2} x_e}_{\text{n. of edges in } E_2} - 1 \\[2mm]
0 \le x_e \le 1 \text{ integer} \,, \, \forall e \in E
\end{cases}
$$

⋄ **Exercise 9–28:**

$$\left\{ \begin{array}{c} \max \sum_{e \in E} c_e x_e \\[2mm] \sum_{e \in E} x_e = |V| - 1 \\[2mm] \sum_{e \in E(S)} x_e \leq |S| - 1 \,,\, \forall\, S \subset V \,,\, |S| \geq 3 \\[4mm] \underbrace{\sum_{e \in E_1} w_e x_e}_{\text{weight of edges in } E_1} \leq W \underbrace{\sum_{e \in E_1} x_e}_{\text{n. of edges in } E_1} \\[6mm] \underbrace{\sum_{e \in E_1} x_e}_{\text{n. of edges in } E_1} \geq \left\lceil \frac{|V|-1}{2} \right\rceil \\[6mm] 0 \leq x_e \leq 1 \text{ integer } ,\ \forall e \in E \end{array} \right.$$

◇ **Exercise 9–29:**

$$\left\{ \begin{array}{l} \min \qquad \sum_{e \in E} c_e x_e \\[2mm] \qquad \sum_{e \in \delta(v)} x_e = b_v \,,\, \forall v \in V \\[2mm] \quad 0 \leq x_e \leq 1 \text{ integer}, \ \forall e \in E \end{array} \right.$$

Bibliography

Basic textbooks

C. Papadimitriou, K. Steiglitz, *Combinatorial Optimization*, Prentice Hall, Englewood Cliffs, New Jersey, 1982.

Linear Programming

V. Chvátal, *Linear Programming*, Freeman, New York, 1983.

Integer Linear Programming

G.L. Nemhauser, L.A. Wolsey, *Integer and Combinatorial Optimization*, Wiley, New York, 1988.

Computational Complexity

M.R. Garey, D.S. Johnson, *Computers and Intractability: A Guide to the Theory of NP-Completeness*, Freeman, San Francisco, 1979.

Graph Theory

N. Christofides, *Graph Theory: an Algorithmic Approach*, Wiley, New York, 1978.

Data Structures

A.V. Aho, J.E. Hopcroft, J. Ullman, *The Design and Analysis of Computer Algorithms*, Addison-Wesley, Reading MA, 1974.

T.H. Cormen, C.E. Leiserson, R.L. Rivest, *Introduction to Algorithms*, The MIT Press, London, 1989.

Applications

F.S. Hillier, G.J. Lieberman, *Introduction to Operations Research*, Holden-Day, Oakland, California, 1986.

W.L. Winston, *Operations Research: Applications and Algorithms*, Duxbury Press - PWS Publishers, Boston, 1987.

Made in the USA
Columbia, SC
15 July 2022

63544477R00128